COMPUTER-INTEGRATED MANUFACTURING HANDBOOK

OTHER BOOKS BY V. DANIEL HUNT

Mechatronics: Japan's Newest Threat
Dictionary of Advanced Manufacturing Technology
Robotics Sourcebook
Artificial Intelligence & Expert System Sourcebook
Smart Robots
RobotJustification
Industrial Robotics Handbook

COMPUTER-INTEGRATED MANUFACTURING HANDBOOK

V. Daniel Hunt

CHAPMAN and HALL
New York   London

First published 1989
by Chapman and Hall
29 West 35th Street, New York, NY 10001

Published in Great Britain by
Chapman and Hall, Ltd.
11 New Fetter Lane, London EC4P 4EE

© 1989 by Chapman and Hall

Printed in the United States of America

Library of Congress Cataloging in Publication Data

Hunt, V. Daniel.
  Computer-integrated manufacturing handbook.

  (Chapman and Hall advanced industrial technology series)
  Bibliography: p.
  1. Computer integrated manufacturing systems—
Handbooks, manuals, etc.   I. Title.   II. Series.
TS155.6.H84   1988   670.42'7   88-2575
ISBN 0-412-01651-6

British Library Cataloguing in Publication Data

Hunt, V. Daniel
  Computer-integrated manufacturing handbook.
  1. Manufacture. Applications of computer
  systems
  I. Title
  670.42'7
  ISBN 0-412-01651-6

Contents

Preface

Manufacturing has entered the early stages of a revolutionary period caused by the convergence of three powerful trends:

- The rapid advancement and spread of manufacturing capabilities worldwide has created intense competition on a global scale.
- The emergence of advanced manufacturing technologies is dramatically changing both the products and processes of modern manufacturing.
- Changes in traditional management and labor practices, organizational structures, and decision-making criteria represent new sources of competitiveness and introduce new strategic opportunities.

These trends are interrelated and their effects are already being felt by the U.S. manufacturing community. Future competitiveness for manufacturers worldwide will depend on their response to these trends.

Based on the recent performance of U.S. manufacturers, efforts to respond to the challenges posed by new competition, technology, and managerial opportunities have been slow and inadequate. Domestic markets that were once secure have been assailed by a growing number of foreign competitors producing high quality goods at low prices. In a number of areas, such as employment, capacity utilization, research and development expenditures, and capital investment, trends in U.S. manufacturing over the last decade have been unfavorable or have not kept pace with major foreign competitors, such as Japan. There is substantial evidence that many U.S. manufacturers have neglected the manufacturing function, have overemphasized product development at the expense of process improvements, and have not begun to make the adjustments that will be necessary to be competitive.

These adjustments represent fundamental changes in the way U.S. manufacturers perceive their competitive advantages, devise competitive strategies, and manage and organize their operations. One response that is beginning to gather momentum among U.S. manufacturers is the implementation of computer-integrated manufacturing. Indeed, technology, wisely applied, can improve costs, quality, flexibility, and responsiveness, but the effects of technology on these areas can be complex. Trade-offs between improving flexibility and responsiveness on the one hand and reducing costs on the other will continue; technologies that are poorly applied may not have the effects intended; and many barriers to their smooth operation remain. Effective implementation of new technology demands a clear definition of the business's strategy and a clear understanding of the role of advanced technologies in supporting that strategy. Managers must recognize that many of the perceived advantages of new technology can be achieved with new management techniques, more effective planning, better coordination across corporate functions, efforts to reduce set-up times and speed changeovers, and simplified part designs to enhance producibility. Having made effective operational and organizational changes, the company can eliminate many of the problems that are often associated with the introduction of new technologies. Effective efforts in these areas also should help managers focus new investments on appropriate technology that can produce dramatic benefits.

These required organizational changes, however, will be difficult for many manufacturers to implement. They require creative initiatives from managers, the cooperation and involvement of employees, and major changes in the relationships at every level of the manufacturing corporation. A fundamental cultural and attitudinal shift will be required on the part of both workers and managers. Manufacturing will need to be thought of as a system, with extensive integration, cooperation, and coordination between functions, to achieve competitive goals. Flatter organizational structures are likely to become the norm and traditional hierarchical relationships are likely to fade as the distinctions between managers and workers blur. Workers will have more responsibility and greater job security and be more active participants in the manufacturing system.

Because the successful implementation of this cultural revolution in the factory depends on thousands of individual initiatives, change is likely to be gradual. In many cases, there will be strong resistance from both managers and workers who have a stake in traditional practices and structures. However, as competition in the new environment intensifies and the requirements to maintain competitive advantage with quality manufactured goods become clear, the benefits and the necessity of implementing these changes will be increasingly apparent.

These changes imply that the factory will provide a much different working environment and play a different role in the macroeconomy.

For example, manufacturing will provide fewer job opportunities for unskilled and semiskilled workers, but the jobs that will be created are expected to require greater amounts of skill and training and thus be more challenging and rewarding. Many manufacturers will have sufficient flexibility built into their production processes to be less affected by shifts in demand, which could moderate business cycles substantially.

These and other effects will require that government officials and the general public adjust their image and expectations of manufacturing. Although the technological and managerial changes necessary for future competitiveness will be the responsibility of the private sector, the government can play an important role in encouraging and supporting these private initiatives. Policymakers must recognize the continuing importance of manufacturing, the need for changes to ensure future competitiveness, and the many repercussions government policies have on the ability of U.S. manufacturers to meet competitive challenges. In addition, some specific government activities, in trade, education, research, and defense, will be affected by developments in manufacturing and will need to adapt accordingly.

V. Daniel Hunt
Springfield, Virginia

Acknowledgments

The information in the *Computer-Integrated Manufacturing Handbook* has been compiled from a wide variety of authorities who are specialists in their respective fields.

The following publications were used as the basic technical resources for this book. Portions of these publications may have been used in the book. Those definitions or artwork used have been reproduced with the permission to reprint of the respective publisher.

Computerized Manufacturing Automation: Employment, Education, and the Workplace (Washington, D.C.: U.S. Congress, Office of Technology Assessment, OTA-CIT-235, April 1984).

A Competitive Assessment of the U.S. Flexible Manufacturing Systems Industry, U.S. Department of Commerce, International Trade Administration, July 1985.

A Competitive Assessment of the U.S. Computer-Aided Design and Manufacturing Systems Industry, U.S. Department of Commerce, International Trade Administration, February 1987.

A Competitive Assessment of the U.S. Robotics Industry, U.S. Department of Commerce, International Trade Administration, March 1987.

Toward a New Era in U.S. Manufacturing—The Need for a National Vision, National Research Council, National Academy Press, Washington, D.C., 1986.

Flexible Manufacturing System Handbook, volumes 1 through 4, U.S. Army Tank-Automotive Command, Warren Michigan, Report CSDL-R-1599, February 1983.

Robotics and Automated Manufacturing, Richard C. Dorf, Reston Publishing Company, Reston, Virginia, 1983.

The preparation of a book of this type is dependent upon an excellent staff, and I have been fortunate in this regard. Special thanks are extended to Janet C. Hunt and Donald W. Keehan for research assistance and to Margaret W. Alexander for the word processing of the manuscript.

Part I

System
Fundamentals

Chapter 1

The Need for Computer-Integrated Manufacturing

For U.S. manufacturing, an extended period of world dominance in manufacturing innovation, process engineering, productivity, and market share has ended. Other countries have become leaders in certain industries, the U.S. market is being flooded by manufactured imports, and U.S. manufacturers are faced with relatively low levels of capacity utilization and declining employment. The reasons for this fundamental change are complex. Improved capabilities and competence of foreign manufacturers are partly responsible. Either government interference or the lack of government support has been blamed. Cultural disadvantages are often cited. Many economists explain the relative decline of U.S. manufacturing simply as economic evolution, with the United States moving toward a service economy. These and other factors have been held responsible for the relative decline of U.S. manufacturing, and all are legitimate partial explanations. The truth remains, however, that U.S. manufacturing is not performing as well as that of many foreign competitors and has lost competitiveness in many industries. Regardless of why the environment has changed, the managerial practices, strategies, and organizational designs applied by U.S. manufacturers have not adapted sufficiently to the changed competitive environment, and, consequently, U.S. manufacturing has not been as successful as that of other countries.

The term competitiveness is subject to a variety of definitions. In simplest form, an industry is competitive if the price, quality, and performance of its products equal or exceed that of competitors and provide the combination demanded by customers. International competitiveness is somewhat more complicated because price is heavily influenced by exchange rates, which cannot be controlled by an individual producer. Many economists would claim that the recent high rate of the dollar has been responsible for any lost competitiveness of U.S. manufacturing, and recent adjustments to the dollar will restore competitiveness. This may or may not be true, however, because exchange rates are only one determinant of product price, and price is only one determinant of competitiveness. Price is also determined by production costs, and quality and performance, including innovation, unique or superior design, and reliability, are in many cases more important determinants of competitiveness than price. If U.S. manufacturers can produce high quality goods with less labor, materials, overhead, and inventory than foreign producers, then competitive production can be ensured. These are the areas in which U.S. manufacturers have lagged—improvements in the use of these resources, as well as product quality and performance, are the requirements for improved competitiveness.

These changes in relative manufacturing strength are occurring at the same time that many technological innovations promise to revolutionize products and processes in manufacturing. Just as major technological breakthroughs spurred industrial development in the mid-eighteenth century (steam power, new engine-driven machinery) and the development of the modern factory system in the late nineteenth century (electricity, the telephone, and mass production techniques), current breakthroughs in electronics, materials, and communications are creating another revolution in manufacturing. Just as earlier changes forced new directions in manufacturing management, production strategies, and national policies for maximizing competitiveness, the competitive and technological changes affecting manufacturing today should create new goals, new priorities, and new expectations in U.S. industry. Many manufacturing managers and national policymakers, however, have been slow to recognize the implications of these developments. U.S. manufacturing is in danger of being unprepared to compete in the coming age, a failure that would cause rapid erosion of the nation's manufacturing base.

Effective response to the changes in manufacturing depends on a clear understanding of the new environment. Although specific developments are difficult to predict with certainty and the types of changes will vary tremendously among industries, likely trends can be identified. Competition will continue to increase both at home and abroad. New products will proliferate; many products will have shorter life cycles and development cycles. Some industries will have smaller production volumes, with more product customization and variety. New technologies, especially those based on microprocessors, will optimize control of the production process and offer entirely new capabilities. Fewer production workers and middle managers will be needed, but the remaining jobs will require higher skill, more technical knowledge, and greater responsibility. Managers will need to manage manufacturing as a system, basing decisions on new, nontraditional indicators. Direct labor costs will decrease significantly, and the costs of equipment, materials, distribution, energy, and other overhead will grow in importance. Quality, service, and reliability will receive much more emphasis as determinants of competitive production.

These trends indicate that competition, both international and domestic, will be more intense and that the factors determining competitiveness will differ substantially from past experience. Strategies and priorities designed to enhance competitiveness in the mid-twentieth century will be far less effective in the future. The new manufacturing environment will be sufficiently familiar to permit many firms to continue to use traditional approaches, but these firms will lose market share, profits, and the ability to compete. In the new environment, it will not be sufficient to do the same old things better. Companies will need to adopt new management techniques, organizational structures, and operational procedures to strengthen their international competitiveness. Government policies must also ensure that U.S. manufacturers receive the infrastructural support they will need to compete effectively.

A Historical Perspective on U.S. Manufacturing

For much of the twentieth century, U.S. manufacturers were unchallenged in an environment in which conservative approaches to both process technology and managerial techniques produced successful results. Foreign competition was minimal, the vast domestic market encouraged product standardization and economies of scale, and the preeminence of Yankee ingenuity was unchallenged. Companies modified strategies and processes in minor ways in response to shifting economic circumstances, but mostly the system worked and they had little incentive to change. The relative stability of the manufacturing environment was unsustainable, however; a series of changes has gradually converted the traditional strategies to handicaps.

One change has been the way companies justify new investment in manufacturing. During the 1950s and 1960s, the emphasis in manufacturing was on providing substantial additional plant capacity that was needed just to keep up with market growth. The addition of capacity provided the opportunity to incorporate process improvements that otherwise were rarely implemented. Beginning in the early 1970s, the rate of growth slowed (Table 1–1), in many cases eliminating the need

Table 1-1. U.S. Manufacturing Output[a]

Average Annual Percentage Change

Period	Total	Durable Goods	Nondurable Goods	As a Percentage of Total Output[b] (average)
1950-1983	3.1	3.0	3.1	24.4
1950-1973	4.0	4.0	4.0	24.6
1973-1983	0.9	0.7	1.1	24.1
Slowdown	3.1	3.3	2.9	0.5

[a]Gross product originating in manufacturing in constant dollars.
[b]Gross national product in constant dollars.

Source: U.S. Bureau of Labor Statistics, 1985.

for additional capacity. Companies needed to develop new justifications for reinvestment in manufacturing, which many have been slow to do.

Another major change in the manufacturing environment was in the process of developing and implementing new innovations. The first Industrial Revolution in the 1800s produced a series of significant innovations in process and product technologies that represented an integration of several types of technologies. In contrast, during the early to mid 1900s, manufacturers, except perhaps for electronics and chemicals manufacturers, increasingly refined proven technologies rather than developing and integrating new and diverse technologies to accomplish, or even eliminate, traditional tasks. This apparent trend toward a more stable, conservative approach to process technology in a broad range of U.S. industries combined with a variety of other factors—such as changing labor demographics, higher energy prices, and lower expenditures on research and development—to cause a shift toward more modest improvements in productivity.

U.S. industries in which new technology did seem to offer great potential focused predominantly on product engineering at the expense of process engineering. (The semiconductor, chemical, and biotechnology industries are exceptions since most of the breakthroughs in their products depend on breakthroughs in process capabilities.) Since manufacturers had their hands full simply adding capacity of a known type, they saw no pressing need to add new process technologies at the same time. Consequently, many U.S. firms spent incremental dollars on product technology and very little on new process technology. Generally speaking, U.S. manufacturers left process development to equipment suppliers and allowed their own skills at such development—and its link with product technologies and product quality—to decline.

The Current Role of the Manufacturing Function

These historic trends illustrate aspects of the manufacturing environment that have shaped the strategies of U.S. managers. For these and a variety of other reasons rooted in the history of industrial development, many managers have focused on increasing the productivity of the manufacturing function, emphasizing production volume instead of product quality and reliability and process development. They believe that manufacturing, at best, can simply provide adequate support for competitive advantages in marketing or design engineering. It is true that many firms, particularly those in Fortune 500, do enjoy substantial advantages in manufacturing owing to economies of scale and degrees of specialization that they have been able to achieve as large organizations. Generally, however, the charge to manufacturing, even in these companies, has been "Make the product—without any surprises."

The traditional view in many U.S. firms is that manufacturing is a problem that can be solved with a given process at a given time. That process is then operated efficiently, with little incremental upgrading, until a significant improvement or new technology is implemented by competitors. This command-and-control view of manufacturing is based on the premise that smart people should be able to determine the optimal solution (process) for handling the tasks of the manufacturing function and then control the process and organization for maximum stability and efficiency until some external event forces change. Since the time between changes varies, the repercussions of this view may not be readily apparent. The key point is that it is a reactive view that overlooks the potential contributions of the manufacturing function to overall competitiveness.

Such an approach can erode the strength and competitive advantage provided by manufacturing. Quality, reliability, and delivery problems get blamed on manufacturing: The plan is assumed to be good, so the people in production must have failed to deliver. The organization increasingly refines the detailed measures of manufacturing in the process of removing degrees of freedom. Scientific management techniques were developed to measure, predict, and control all the aspects of production in an effort to limit change, or at least eliminate surprises, and achieve maximum productivity. Advances in production planning, project evaluation, and operations research offered new tools for maintaining stability and increasing productivity. The introduction of computers and manufacturing information systems in the late 1960s and early 1970s was hailed as finally giving manufacturing a tool that could be used to pursue the command-and-control approach to operations. Although designed to ensure stability in daily operations, these detailed measurements and sophisticated control tools too often can become ends in themselves and impediments to process changes.

The consequence of this approach to manufacturing has been increased tuning and refining of a set of resources that were outdated and increasingly inappropriate. The individual firm often slipped into a debilitating spiral: Additional investment was withheld because the current investment was not performing as expected; those operating the current investment simply tried to minimize the problem in the near term rather than looking for long-term solutions they knew would not be approved and supported.

Recent Performance of U.S. Manufacturing

The repercussions of this command-and-control approach, with its reactive nature and short-term focus, are not difficult to find. The United States has sustained a steady erosion of competitiveness and overall manufacturing strength over the past two decades that must be attributed at least partly to deficiencies in standard management practices in manufacturing. Individual companies have adapted to the new environment and fared well, but overall the picture has been bleak. Declining growth trends in manufacturing output have already been cited. Other indicators include:

• Growth in manufacturing productivity (output per man-hour) in the United States during the past 25 years has been among the lowest in the industrial world (Table 1-2). Although manufacturing productivity

Table 1-2.   Output per Hour in Manufacturing, Average Annual Percent Change

Country	1960–1973	1973–1983
Canada	4.7	1.8
France	6.5	4.6
West Germany	5.7	3.7
Japan	10.5	6.8
United Kingdom	4.3	2.4
United States	3.4	1.8

Source: U.S. Bureau of Labor Statistics, 1985.

Table 1-3. Foreign Labor Cost Components
in Relation to U.S. Producers[a]

Country	1970	1975	1980	1981	1982	1983	1984
Average Hourly Compensation							
France	41	72	92	75	68	63	56
West Germany	56	97	125	97	90	85	75
Italy	42	73	81	68	63	62	58
Japan	24	48	57	57	49	50	50
Korea	N.A.	6	11	10	10	10	10
United Kingdom	36	51	75	65	58	51	46
United States	100	100	100	100	100	100	100
Output per Hour							
France	65	70	82	81	85	86	87
West Germany	66	71	79	78	79	79	78
Italy	56	60	70	70	71	68	69
Japan	44	52	72	74	79	79	84
Korea	N.A.	15	17	18	18	17	18
United Kingdom	41	43	42	43	44	45	44
United States	100	100	100	100	100	100	100
Unit Labor Costs							
France	63	103	112	92	80	74	65
West Germany	85	136	157	123	114	107	95
Italy	75	123	115	96	89	91	84
Japan	53	92	79	77	62	63	60
Korea	N.A.	39	63	57	59	60	56
United Kingdom	86	120	177	151	132	115	105
United States	100	100	100	100	100	100	100

Note: N.A. indicates data not available.
[a]Based on U.S. dollar values derived from average annual exchange
rates.

Source: Data Resources, Inc., 1985.

in this country remains the world's highest, it has been virtually
equaled in recent years by Japan, France, and West Germany. Based
on average hourly compensation and output per hour, unit labor
costs in U.S. manufacturing have been higher than those of our major
competitors (Table 1-3).
* In contrast to the growth in manufacturing trade surpluses enjoyed
  by Japan and West Germany, U.S. performance over the past 15 years
  has been highly erratic, with deficits of $31 billion in 1983, $87 billion
  in 1984, and more than $100 billion in 1985 (Table 1-4).
* By 1984, manufacturing output was 8 percent above the previous peak
  in 1979. Defense production, however, accounted for more than 40
  percent of that increase; nondefense output has risen less than 1
  percent annually since 1979, compared with 3.5 percent annually from
  1973 to 1979.
* Recent employment trends have been unfavorable in most durable
  goods manufacturing industries, particularly import-competing industries (Figure 1-1).

Table 1-4. Trade Balance in Manufacturing, Billions of U.S. Dollars

Year	United States	Japan	West Germany
1970	3.4	12.5	13.3
1971	0.0	17.1	15.0
1972	-4.0	20.3	17.7
1973	-0.3	23.3	28.7
1974	8.3	38.0	42.4
1975	19.9	41.7	38.7
1976	12.5	51.2	42.1
1977	3.6	63.0	46.9
1978	-5.8	74.2	53.5
1979	4.5	72.0	59.2
1980	18.8	93.7	63.1
1981	11.8	115.6	61.7
1982	-4.3	104.0	67.5
1983	-31.0	110.3	58.7
1984	-87.4	127.9	60.5
1985	-107.5	107.7	59.5

Source: U.S. Department of Commerce, Bureau of Economic Analysis, 1985, and International Trade Administration, 1986.

• Capital investment as a percentage of output in U.S. manufacturing has increased slightly over the past ten years (Table 1–5), but the composition of investment has tended to neglect traditional industries and new factory construction. Although U.S. manufacturing investment has shown some improvement, it has continued to be below that in other countries.

• A major reason for the level and types of investment in manufacturing, cited by other reports on U.S. manufacturing, is the high cost of investment capital. The cost of capital in this country is far higher than in other nations, and the return on manufacturing assets has not kept pace with the return on financial instruments (Table 1–6, Figure 1–2). In addition to the obvious impact this differential has on investment costs, lower capital costs and different sources of capital

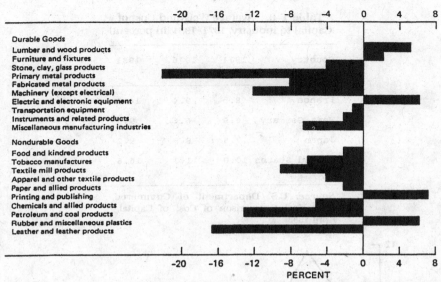

Figure 1-1. Change in manufacturing employment, selected industries (July 1981 to May 1984). Based on seasonally adjusted data for private manufacturing; includes only payroll employees. *Source:* U.S. Department of Labor, Bureau of Labor Statistics, establishment survey data.

allow some foreign competitors to succeed with much lower rates of after-tax profit on sales than U.S. firms (1 to 2 percent versus 5 to 6 percent for U.S. firms). This difference effectively provides extra funds for capital investment or research and development expenditures.

There are a myriad of explanations for these troubling trends in U.S. manufacturing. Management, labor, and government all share responsi-

Table 1-5. Capital Investment as Percentage of Output[a] in Manufacturing, Selected Countries, 1965–1982

Period	France	West Germany	Japan	United Kingdom	United States
1965–1982	15.1	12.8[b]	21.2	13.6	10.5
1965–1973	16.5	14.3	25.3	14.3	10.0
1974–1982	13.6	11.2[c]	17.1	13.0	11.1

[a]Fixed capital and output measured in constant dollars.
[b]1965–1981.
[c]1974–1981.

Source: U.S. Bureau of Labor Statistics, 1985.

Table 1-6. Average Weighted Cost of
Capital to Industry, 1971–1981 (in percent)

Country	1971	1976	1981
France	8.5	9.4	14.3
West Germany	6.9	6.6	9.5
Japan	7.3	8.9	9.2
United States	10.0	11.3	16.6

Source: U.S. Department of Commerce,
"Historical Comparison of Cost of Capital,"
April 1983.

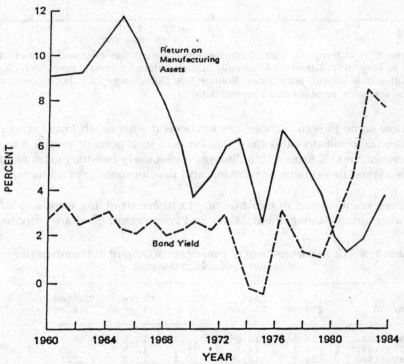

Figure 1-2. U.S. real return on manufacturing assets and industrial bond yield, 1960–1983.

Sources: Quarterly Financial Reports of Mining, Manufacturing and Trade Corporations,
Federal Trade Commission, 1960–1961; Census Bureau, 1981–1984; inflation data from
"Economic Report of the President," 1984; "Moody's Industrial Bond Yield" from *Survey
of Current Business* (July 1984); and *Business Statistics,* 1979 and 1982.

bility. Macroeconomic factors such as domestic interest rates, exchange rates, the availability and cost of labor, foreign and domestic trade policies, and the constant seesaw of business cycles all have had an impact. Uncertainty about government spending, tax, and regulatory policies, and changes in the relative attractiveness of nontechnological (even nonmanufacturing) investments have deterred risky investments in new process technologies and bred caution in managers. Pressure from stockholders, standard financial evaluation procedures, and the disruptive effect that new technology can have on short-term operational efficiency also have caused managers to give priority to maximizing returns on existing assets.

Other Evidence of a Changed Manufacturing Environment

Because of the diversity of the manufacturing sector and the factors affecting manufacturing output and trade, there is little agreement among economists and policymakers that U.S. manufacturing is losing competitiveness. Some authors have used economic data to demonstrate that U.S. manufacturing remains generally strong despite the problems of a few industries. Many reports have addressed the issue by using macroeconomic data, but they have had little impact on either policymakers or the general public.

Statistics on the manufacturing sector tend to be inconclusive because of the complex, transient economic factors that affect the data. Other indicators, however, show that at least some U.S. companies have perceived both eroding competitiveness and a basic change in the nature of the manufacturing environment. These data tend to be anecdotal and industry-specific and can be illustrated by a few examples.

• Through improved management, changed work rules, large investments in automation, and a variety of other measures, the three major U.S. automobile manufacturers have reduced their break-even volume for domestic production by more than 30 percent since 1980. Despite this dramatic improvement, estimates of the cost advantage of Japanese producers have grown from $1,000 to $2,000 per car in 1979–1980 to $2,000 to $2,600 in 1985. Based on consumer surveys, U.S. cars also have lagged behind Japanese makes in quality. U.S. producers have launched new projects—General Motors' Saturn, Chrysler's Liberty, and Ford's Alpha—to eliminate these gaps by rethinking management concepts, employee relations and compensation, and technology. All three companies also are aggressively pursuing joint ventures with foreign producers and captive imports of finished vehicles and parts from several countries to offset the cost disadvantage of domestic production. The companies' approaches differ, but these programs clearly indicate that senior managers in the U.S. automobile industry recognize both the shortcomings of traditional practices and the opportunities that new technologies and new management approaches will provide. Although these efforts may solve current competition problems with Japan, in many cases they will not become operational until about 1990. By that time, other disadvantages and new competitors may have emerged that U.S. firms will be ill-equipped to address.

• Companies in other industries also have aggressively pursued a strong competitive position only to be confronted by intensified competition. Black and Decker Corporation, for example, has devoted significant effort to reducing costs and increasing efficiency by pursuing new investments, increasing automation, reducing its work force, and standardizing parts and product lines across its international operations. Despite these long-term efforts, the company faced growing competition in the world power tool market from Makita Electric Works, Ltd., of Japan and lost a significant part of its market share. Since 1980, Black and Decker has spent $80 million on plant modernization, cut its work force by 40 percent, and adopted new manufacturing practices. The company has regained a 20 percent share of the world market in power tools at the cost of reduced profits resulting from price pressures from the Japanese company. The efforts by Black and Decker indicate the kinds of commitment that are absolutely necessary to maintain a worldwide competitive position.

• The Japanese are not always the prime competitors. Chaparral Steel Company, a minimill operation based in Midlothian, Texas, figures that if it can produce steel at a labor cost per ton no higher than the per ton cost of shipping steel to this country from Korea, it can beat Korean producers. In achieving this goal, the company has invested in some of the most modern steel plants in the world and can produce steel using 1.8 man-hours per ton, compared with 2.3 for the Koreans and 6+ for integrated U.S. producers. Although its capacity and range of product are more focused than that of large integrated producers, Chaparral illustrates two important aspects of the new manufacturing environment. First, the company's experience (and that of other minimill operations, such as Nucor Corporation) has shown that U.S. producers can be world leaders and can pose more of a threat to traditional U.S. manufacturers, albeit in a relatively narrow product line, than foreign competitors. Second, traditional competitive targets, such as matching the production costs of competitors, may not be enough to ensure long-term competitiveness; other targets, such as the shipping costs used by Chaparral, may need to be considered.

• A final example comes from the computer disk drive industry. Floppy disk drives are used in lower end home computers and personal computers, whereas rigid disk drives are used most often in advanced personal computers and engineering work station products. The disk drive industry was created by U.S. electronics firms from technology developed by International Business Machines and Control Data Corporation. Several smaller firms entered the field in the mid-1970s and quickly grew to substantial size. In recent years, however, the status of U.S. disk drive manufacturers has changed considerably. The leading U.S. maker of floppy disk drives in 1980 was Shugart Associates. Xerox Corporation, the owner of Shugart, has since announced the closing of the unit as a disk drive manufacturer. In 1984, more than 20 Japanese firms manufactured floppy disk drives; no U.S. manufacturers did so. While the United States retains a strong position in rigid disk drives, the Japanese are likely to dominate the next-generation technology, optical disk drives. Developments in this industry show that being the first to market, even with high technology products, is not a long-term advantage. Constant improvement in both products and processes is needed to ensure survival.

As these examples illustrate, pervasive and potentially damaging change is overtaking U.S. manufacturing across the spectrum of industries from traditional to "high tech." Industries as diverse as motorcycles, consumer electronics, and semiconductor memories also have been subject to lost leadership and declining market shares. Many firms recognize the change and are responding, though often in limited ways. Many more do not recognize the problem or think that it does not apply to them or their industries. Still others attributed their difficulties to the recent high value of the dollar and are looking forward to the benefits of the recent dollar depreciation. Factors such as interest and exchange rates and unfair foreign competition do have significant effects on industrial health. Unfavorable trends in these areas, however, provide easy scapegoats and disguise other important factors that are changing the manufacturing world. A majority of U.S. manufacturers need to recognize that lowering the cost of the dollar in international currency markets, while important, will not solve all their competitive problems. The price elasticities of many important U.S. imports and exports will determine the long-term effect of the recent decline in the dollar. Although some U.S. commodity exports, such as timber, coal, and some agricultural goods, are likely to increase as the dollar declines, exports of capital goods and major imports of items such as machine tools, automobiles, and consumer electronics may change little as the dollar's value changes, at least in the near term. Many consumers continue to prefer foreign goods because of perceived quality and reliability advantages over their

U.S. counterparts. Furthermore, many foreign companies in a range of industries have advantages in production costs that permit them to offset even unexpectedly large devaluations of the dollar by limiting price increases in the U.S. market.

More U.S. firms need to join the minority that recognize the challenges emerging in manufacturing and are devoting resources to meet them. Although competitive challenges are spreading to more and more products and industries, too few companies are making the essential commitment to competitive manufacturing operations in the United States. The rising competition from previously weak or nonexistent sources is prompting a response, but it is insufficient. The initial, and natural, reaction is to do everything better. Redoubled efforts are nearly always beneficial; it is a rare company that does not have room to improve. Doing things better than yesterday or better than competitors today, however, will not necessarily ensure long-term competitiveness.

Another response has been to move production facilities offshore, through foreign direct investment, outsourcing, joint ventures with foreign producers, or other mechanisms. While such arrangements have clear, short-term advantages in terms of foreign market penetration and labor cost containment, the long-term repercussions of offshore production strategies are not clear. In some industries, firms must move constantly in search of even lower wage rates; in others, host countries insist on domestic content, technology transfers, and domestic equity positions that lead to independent, competitive production capabilities. Factors vary across industries, and some firms in labor-intensive industries may have no choice but to move production offshore or purchase components or products from abroad. As technological developments yield effective alternatives to offshore production and conditions for foreign direct investment become more stringent, a better understanding is needed about the effects of offshore production strategies on the long-term interests of individual firms and the domestic industrial base.

Another response from U.S. manufacturers has been based on the widely held idea that technology will solve the problem. Advanced manufacturing technology, such as computer-integrated manufacturing, can provide dramatic improvements in efficiency, but only if the groundwork is laid. The benefits of new technology will not be fully achieved if the organizational structure and decision-making process are not changed to take advantage of available system information, if the work force is not prepared for the changes brought by the technology, and if potential bottlenecks created by automating some operations but not others are not foreseen and avoided. Many companies that have powerful computer-aided design (CAD) systems, for example, are using them as little more than electronic drafting boards, negating many of the capabilities that CAD provides because appropriate adjustments in the organization have not been made. Managers need to understand that technology is both a tool for responding to competitive challenges and a factor causing change in manufacturing.

Recent economic data and the experiences of specific industries suggest that a strong case can be made that U.S. manufacturers, with the exception of a handful of enlightened companies, are not responding adequately or entirely appropriately to new competitive challenges, even as those challenges intensify. The first corrective step is to convince managers that they face a manufacturing problem that new technology, offshore production, changes in exchange rates, and redoubled efforts cannot resolve. The next step is to indicate the kinds of changes in manufacturing organizations that will be needed to maintain long-term competitiveness.

The changes needed can be described broadly as a shift from the traditional management goal of maximizing stability, productivity, and return on investment in the short term to the new goal of maximizing adaptability to a rapidly changing market, with long-term competitiveness as the first priority. A number of authors have detailed the changes that are necessary in the management of the manufacturing function. Hayes and Wheelwright, for example, describe the needed changes as a shift from an "externally neutral" role for the manufacturing function, in which the firm only seeks to match the process capabilities of its competitors, to an "externally supportive" role, in which process improvements are continually sought and implemented in an effort to maintain a lead over competitors, and manufacturing is viewed as a significant contributor to the firm's competitive advantage. This shift cannot be made overnight, and it is far too easy to backslide once a new plateau is reached. The shift requires changes in organizational structure and decision-making processes, and it demands new skills: Managers must learn to manage change.

Stakes for the U.S. Economy

Because manufacturing remains crucial to national economic and defense interests, the repercussions of declining competitiveness could be devastating. Many economists argue that continued erosion of the domestic industrial base is limited because manufactured goods are the major component in international trade. The United States will remain a major manufacturer because world markets may not tolerate a constant large U.S. trade deficit. Exchange rates will adjust to ensure that the United States can export manufactured goods. Alternatively, the United

States will suffer a recession that will dampen demand for imports and alleviate the trade deficit. Recent historical evidence for this argument, however, is ambiguous at best: the United States managed only small surpluses in manufactured goods during the late 1970s, when the dollar was relatively weak, and had a small deficit in the recession year of 1982 (see Table 1–4). Particularly because exchange rates increasingly react more to financial flows than to flow of goods, the sustained process of devaluation of the dollar necessary to maintain the competitiveness of U.S. manufacturers would be difficult to accomplish both economically and politically.

Recessions and shifts in currency value can be painful ways for the nation to reach equilibrium in its manufacturing trade. An alternative is for U.S. manufacturers to implement the organizational, managerial, and technical changes necessary to maintain a strong manufacturing sector. Competitiveness would be based on leadership in product performance and quality rather than on a declining exchange rate. The resulting strength would provide the basis for continued economic growth and provide crucial advantages in areas of national importance, such as:

- Defense. Counter to conventional ideas that a strong industrial production base is necessary to meet U.S. defense commitments, some economists have argued that these commitments can be met without broad support for manufacturing. Although it may be possible to meet them through selective policies designed to support specific defense industrial production instead of entire industries, such an approach would be inadvisable for two major reasons. The first is that it would not provide the productive capacity needed for surges or mobilization in the event of a prolonged conventional engagement. The second reason is that selective policies would hinder the ability of defense contractors to maintain broad technological superiority. This, in turn, would limit their flexibility in response to new defense needs. Production capacity and the technological level of weapons systems are closely linked. Advanced weapons that maintain the qualitative advantage built into the U.S. defense posture require complex manufacturing processes and advanced production equipment, which in turn require broad-based manufacturing capacity. Both the weapons and the production processes are most effectively developed and implemented in the broad context of a healthy manufacturing sector.
- Living standards. Although an absolute decline in the manufacturing base might be countered in the short term by growth in the service sector, services are unlikely to be able to absorb a large percentage of unemployed manufacturing workers at their customary level of wages and benefits. Economists disagree about the validity of projections of

a shrinking middle class, but declining manufacturing employment would certainly have a large impact on total wage and benefit packages. The increased competition for jobs in services, as well as the likely increase in competition among firms in that sector, would moderate wage growth in services.

International competition in services also can be expected to intensify and moderate wage levels. Apart from the effect of this competition on wages, sufficient growth in services is not at all ensured because many services are tied to manufacturing; if manufacturing decays, these services will decline, too. Furthermore, there is no guarantee that the United States can maintain a comparative advantage and large trade surplus in services that would be necessary to pay for manufactured goods. It is not at all clear that the nation's long-term economic strength lies in services or that potential strength in services is greater than its potential strength in manufacturing. Given these considerations, the extent to which service industries can absorb workers displaced from a declining manufacturing sector, and continue to drive overall economic growth remains in doubt.

A technologically advanced manufacturing sector also would result in displacement of workers, but a competitive, dynamic economy should be much more successful at creating new jobs. The development of new products, technologies, and support needs would create whole new industries with job opportunities that would be unlikely to develop in a stagnating manufacturing sector.

National economic and political goals in the domestic economy, and regional concentration of manufacturing activity creates the potential for economic disruption from a declining manufacturing sector that would be disproportional to its share of the gross national product. The decline of whole communities dependent on a single factor is, of course, not a new phenomenon, but past experience has clearly shown that the necessary adjustments are difficult and costly. Services in those regions and communities tend to depend on manufacturing and are ill-equipped to provide employment and generate income in the face of a declining industrial infrastructure. The decline of U.S. manufacturing would have a severe adverse effect on these regions, and the national policies that would be necessary to support them would be politically difficult to enact. These patterns of regional strength and weakness serve to exacerbate the national economic dilemmas posed by a decline in U.S. manufacturing.

On an international scale, the sheer size of the domestic market is a major driver of economic development, competition, and continued advances on a global scale as a growing number of foreign manufacturers compete for a share of the U.S. market. A declining ability of the United States to supply its own manufactured goods, however, would fundamentally change the relationship between this country and foreign manufacturers. Domestic companies would have less revenue and incentive to pursue strong research and development programs, leading to less innovation and invention and fewer patents. The lack of manufactured goods to trade and of manufacturing income to purchase foreign goods would reduce the bargaining position of U.S. producers and the attractiveness of the U.S. market.

These points illustrate the importance of a strong manufacturing sector and the danger of considering the demise of U.S. manufacturing in purely economic terms. Clearly, each individual industry need not survive, but manufacturing as an economic activity is too important to let decay. Changes in management, process technology, corporate organization, worker training, motivation, and involvement, and government policies are necessary to ensure that resources flow to manufacturing. Changing traditional ideas about education, the role of workers, investment in research, development, and innovation, and overall attitudes toward manufacturing will require input and active participation from a variety of sources. The transition will not be painless. Job displacement, plant closures, and changing industrial patterns will be the norm, as they always have been. But these events will take place in a dynamic economy and, therefore, will be accepted and resolved as smoothly as possible. The result will be a competitive manufacturing sector, far different from today's, which will be a leader in the new era in manufacturing.

The Role of CIM

The recent performance of U.S. manufacturing suggests that relatively few firms have begun to make the changes necessary to compete in the manufacturing environment of the future. As competition intensifies and technological change affects both products and processes, standard practice will become less and less effective. In many cases, success in the marketplace will require reassessment of total business strategy and the tools used to pursue it. Functions such as purchasing, marketing, and distribution will need to adapt, but the primary need in many industries will be suitable design and production strategies. In fact, the effectiveness of the design and production functions in supporting overall business strategy may prove to be the major determinant of the competitiveness of U.S. industry.

Effectiveness is a relative term that should not necessarily be equated with the most advanced design and production technologies. Selecting the proper technologies is a major management challenge that will become more complex as technology advances. The selection will depend on factors such as the capabilities of the technology, the type and number of products being manufactured, the relative costs of production, the abilities of competitors, and the long-term objectives of the company. Advances in computer-integrated manufacturing systems (CIM) may be more valuable to a batch manufacturer with frequent modifications in machining requirements than to a mass producer of standard parts with infrequent design changes. Each manufacturer, even each plant, must assess its objectives and employ technology accordingly.

Management also must implement technology effectively. Effective implementation requires recognition of the effects that new technologies can have on total operations. Material handling systems and automated assembly systems, for instance, may be installed to reduce direct labor inputs, but they also can result in lower inventory costs, a streamlined production process, and new part designs. Such systems, however, may create a need for additional computer hardware and programming talent as well as specialized engineers, so total labor costs may not change much. Managers also must understand that needs and capabilities may change rapidly. None of these technologies can be one-time investments.

Instead of focusing strictly on the physical capabilities of the technology, this book emphasizes issues in using advanced technologies in design and production processes; issues that may represent radical departures from traditional manufacturing experience. Some companies are already confronting these issues in their efforts to modernize production facilities. These U.S. manufacturers are well into the early stages of a long-term technological revolution in manufacturing. Their early experiences give strong indications of the direction that manufacturing management, strategy, and competition will take in the future.

New technologies will permit manufacturers to implement strategies that previously were, at best, ideal objectives subject to a range of compromises. These strategies will vary by company, but several objectives will be common to all. Every manufacturer will continue to strive for more efficient use of total resources. Under this overriding goal, key objectives can be summarized as (1) more rapid response both to market changes and to changing consumer demands in terms of product features, availability, quality, and price; (2) improved flexibility and adaptability; and (3) low costs and high quality in design and production. More specific categories could be added, but these three broad objectives are likely to be major elements of successful competitive strategies. Clearly, new technologies cannot achieve these goals. Great strides can be made using new management approaches, more flexible work rules, and organizational changes. In fact, these factors are absolutely crucial to improving manufacturing effectiveness, particularly since new technologies are not likely to achieve optimal results without modifications in the human resource aspects of the organization. Managerial and organizational changes are necessary conditions for improved manufacturing competitiveness—new technologies can increase those benefits exponentially.

Responsiveness to Technological Change

A manufacturer's response to change varies with specific circumstances, such as the structure of the industry, the particular product line, the market in which the product is sold, and the firm's competitive emphasis. Because of the many permutations of external factors and possible responses, the important requirement is not a strategy for every contingency but the ability to pursue a range of strategies aimed at particular combinations of circumstances. Much improvement in responsiveness can be achieved by reevaluating the company's operations, particulary in design, engineering, and manufacturing, to determine handicaps, improve functional cooperation, and strengthen common goals. Cooperation and even integration of the many functions in the entire manufacturing system will need to be pursued aggressively to achieve many of the necessary improvements in producibility, productivity, quality, and responsiveness. In the future, these efforts will be strengthened by the capabilities embodied in advanced manufacturing technologies, such as CIM. When combined with effective organizational changes, advanced manufacturing technologies will be powerful tools for achieving enhanced responsiveness to many external factors that affect design and production.

The technologies that can help maximize responsiveness will vary among firms and product types, and an enormous amount of tailoring will be involved. A major consideration will be the degree to which the firms' competitive strategy depends on price leadership or product differentiation. A low price strategy implies the ability to offer the combination of performance and quality demanded by customers at the lowest possible price, yet still respond to demand changes, variations in input availability and relative costs, and changes in competitors' capabilities. A product differentiation strategy, on the other hand, aims to supply a range of price and performance options that covers most consumer demands. A firm with a low price strategy may benefit more from improved production, material handling, and inspection technologies, while a product differentiation strategy may demand more emphasis on computer-aided design, flexible manufacturing systems, and new materials and processing techniques.

For many manufacturers, a CAD system is a necessary first step in improving responsiveness. A CAD system that includes a comprehensive CAD data base and modeling capabilities can provide significant benefits beyond such simple use as an electronic drafting board. It can speed changes in product design to meet market demand by improving the productivity of design, manufacturing, and process engineers and ensuring that new designs are readily producible. When coupled with an effective group technology (GT) data base, such a system permits parts to be designed and produced rapidly and effectively with minimum changes in tooling, numerical control programs, and material requirements because new parts will resemble current parts as much as possible. Of course, designing for producibility is a crucial goal regardless of the technology used, but a CAD system can improve the designer's ability to simplify new and existing part designs, improving manufacturability and reducing the need for complex multiaxis machining. The benefits in terms of design costs, design to production times, and quality can far outweigh investment costs. The full benefits will not be achieved, however, without vigorous efforts by management to train engineers and technicians, encourage acceptance, and change the corporate culture, support systems, and business philosophy in a way that will allow the capabilities of the CAD system to permeate the company. Improved design response does no good if new designs are allowed to languish on the factory floor.

Although some type of CAD system and GT data base will be a major factor in most firms' efforts to maintain competitive response times, the types of process technology that offer substantial benefits will vary tremendously among firms depending on the types, variety, and volume of parts produced. With batch parts, for example, numerically controlled (NC) machining centers and turning centers with cutting tool drift sensors, self-correcting mechanisms, and laser-based inspection equipment should provide significantly better quality, reduced scrap and rework, and shorter response times. The manufacturer could be more confident that each part was within specifications and, therefore, could respond more quickly and effectively to changing consumer demands.

A computer-integrated manufacturing system (CIM) would take these benefits a step further. A CIM can perform a number of operations on a limited variety of parts within a family (for example, parts having a fundamental aspect in common) without the need for manual handling of the parts between operations. Given this definition, a single complex NC machining center conceivably could be described as a CIM. More sophisticated CIM systems might combine a number of machining centers with material handling systems and automatic tool and part changers and integrate the required CAD, GT, material, tooling, and NC program data. In the future, automated assembly operations will be included in the CIM installation, eliminating much of the required human intervention, but this development depends on progress in a number of technological areas. Current CIM installations cover a range of equipment combinations, complexity, and capabilities, which indicates that manufacturers' production operations must be evaluated carefully in developing a CIM that is effective and appropriate for production needs. For many applications, responsiveness through rapid set-up and delivery is as beneficial as the flexibility offered by a CIM. In many other cases, however, CIM systems provide more complexity, capabilities, and costs than are really necessary for the firm's production, and alternate approaches, such as design simplification or hard automation, would be more appropriate.

Although most mass production industries may not need the capabilities promised by a CIM, computer-integrated manufacturing (CIM) systems will find applications in all types of manufacturing regardless of the process machinery on the factory floor. CIM refers basically to the data-handling capabilities of the manufacturer. It is a sophisticated system for gathering, tracking, processing, and routing information that links purchasing, distribution, marketing, and financial data with design, engineering, and manufacturing data to expand and speed the knowledge available to employees and managers. CIM systems will use interactive data bases and hierarchical control systems coupled to advanced CAD systems, modeling and simulation systems, computer-aided engineering (CAE) systems, production process planning systems, computer-integrated manufacturing systems and/or hard automation processes, material handling systems, and automated inspection/quality assurance systems. At a minimum, each of these elements will be necessary for an effective CIM system because each builds and depends on the others. Each also will require substantial investment in hardware, software, and personnel, but the capabilities provided by CIM can be expected to introduce new priorities and new determinants of competitive advantage.

In essence, a complete CIM system will allow teams of design, process, and manufacturing engineers to design products quickly in response to market demand, product innovations, or input price changes. The CAD system, modeling system, CAE systems, GT data base, and other manufacturing data bases will provide feedback on producibility, material requirements and availability, various production process options, costs, and delivery schedules. The design data will then be transferred directly to the other elements of the CIM system, and the new product can be produced within a very short time. For many products, the interaction of the engineers with the CIM system may be the only significant direct human participation in the production process. Such a capability has already been demonstrated, at least for simple metal parts in very low volume, at the National Institute of Standards and Technology Automated

Manufacturing Research Facility, as well as by several private manufacturers. Substantial research will be required, however, to expand and enhance both the hardware and software technology to allow high volume production of complex parts.

Manufacturers in the future are likely to find that the responsiveness provided by CIM is a necessary competitive advantage. By providing virtually immediate information on every aspect of the manufacturing enterprise, CIM will force manufacturers to eliminate delays in design and production, because rapid reaction to new opportunities and changing conditions will be a major competitive factor. These developments will introduce new bases for competition and may change the economic environment in many industries. Competition will be based on management and labor skills, organizational effectiveness, the price and quality of the final product, speed of delivery, and serviceability of the design, including appropriateness of materials used, functionality, ease of repair, and longevity. Since CIM technology will be readily available, market success will depend on proprietary refinements to the CIM system and how well it is used. The manufacturer who can use CIM not only to respond quickly but also to minimize total resource requirements will have a competitive edge.

Flexibility

Responsiveness, as discussed here, refers to a manufacturer's ability to react quickly to changes in external conditions; flexibility is really an extension of that concept to internal factors. In fact, one can distinguish several types of flexibility.

- Process flexibility is the ability to adapt processes to produce different products without major investments in machines or tooling for each product. This type of flexibility is the cornerstone of flexible automation technologies that allow optimal matching of materials to product applications, the ability to use various materials, and the ability to produce a variety of product designs.
- Program flexibility allows process path modifications, adaptive control and self-correction, unattended operations, and back-up capabilities to maintain production even when some part of the process fails. This type of flexibility addresses the need to optimize equipment use and run multiple workshifts.
- Price volume flexibility provides the ability to maintain economic production in a wide range of market conditions resulting from cyclical and seasonal changes in demand. (This type of flexibility concerns ex-

ternal circumstances, but the firms' internal flexibility will determine the success of the response.)

• Innovation flexibility refers to the ability to implement new technologies as they become available. This type of flexibility depends on a modular approach to manufacturing systems integration that is essential to an evolving design and process capability.

Manufacturers have always confronted the problems these various types of flexibility address and to solve them have relied on compromises between production capabilities and costs. The costs of increased flexibility traditionally have included higher inventory, increased tooling and fixturing requirements, lower machine utilization, and increased labor costs. By reducing some of these costs through advanced manufacturing technologies, more producible part designs, streamlined organization, better functional cooperation, just in time inventory control systems, material requirement planning, and other mechanisms, future manufacturers have the opportunity to reduce the need for many of the traditional cost flexibility compromises. Cost flexibility trade-offs will always exist, however, and the advantages of specific technologies will vary tremendously among industries and product lines.

For many mass production industries, conventional hard automation will remain the most efficient production process. Hard automation—that is, transfer machining lines—is relatively inflexible; it is generally product-specific. Advances in design capabilities, sensors, materials, robotics, material handling systems, and automated inspection technologies should introduce a degree of flexibility into these operations, but efficient production will still tend to depend on economies of scale and product standardization. For traditional batch part manufacturers, the flexible automation technologies embodied in a CIM will change many of the historic cost flexibility compromises. In some applications, CIM systems with flexible fixtures can be expected to reduce set-up times to near zero, allowing smaller and smaller lots produced on demand to become both economically feasible and competitively necessary. In other applications, group technology, designing for producibility, and efforts to speed changeovers will increase flexibility at less cost and with greater effectiveness than elaborate CIM installations.

Although the flexibility of the process equipment on the factory floor will differ between mass producers and batch manufacturers, both will benefit from the flexibility embodied in CIM. The ability to gather and manipulate data in real time as orders are received and products are made will provide a degree of control over the manufacturing process that has not been possible in the past. It is in this context that the issues and benefits of flexibility become particularly relevant to a competitive production strategy.

No manufacturer could afford, and no technology could provide, infinite flexibility, and increasing investments in advanced technologies will not necessarily correlate with increasing flexibility in production. As an extreme example, a machine shop using manual machine tools and expert craftsmen may be more flexible in producing a broad range of parts and may be better able to improvise to produce prototypes than a more modern machine shop using NC machine tools. The manual shop, however, is likely to be less cost effective and slower, have more scrap and rework, and, most importantly, be ill-prepared to take advantage of other computer-based technologies that could improve control over and the effectiveness of the total production process. The NC shop may have a narrower product line, but it is likely to have quicker response times, more consistent tolerances, better repeatability, and greater ability to integrate other computer-based technologies, such as CAD, and use them effectively. Strictly from a production perspective, the manual shop could be described as more flexible, but from a total operations perspective, the NC shop is more flexible. Its potential for introducing new design and production technologies, particularly the data-handling capabilities of CIM, is incomparably greater than that of the manual shop. Neither approach is indisputably correct, however; each is based on value judgments and trade-offs made by the owners in response to their circumstances. This example, although extreme, illustrates the unavoidable cost-capability compromises that will always be confronted in the pursuit of greater production flexibility.

Determination of an optimal level of flexibility for a given plant must include not only the cost effects of different ranges of product mix and quality relative to production capacity, but also the cost effects of fluctuations in demand. The greater the investment in production facilities and the associated fixed costs, the higher the break-even rate of capacity utilization. Consequently, cost savings expected from more capital-intensive production systems at high levels of utilization must be balanced against higher average unit costs as seasonal and cyclical fluctuations reduce average capacity utilization. This dilemma involves price volume flexibility, but the lesson is applicable to all four types of flexibility. Significant planning is required to achieve optimum flexibility; ad hoc programs and investments will be counterproductive.

Although cost-flexibility compromises will continue to apply, the basic flexibility provided by new technology will be much greater than with current NC machining and turning centers or CIM installations. Future computer-integrated manufacturing systems are expected to accommodate a variety of process plans resulting in optimum use of equipment. The system will permit variation in the sequence of operations on the same part, constrained only by the need to drill a hole, for example, before reaming it. This multiple path flexibility will be feasible not only because set-ups will be flexible and essentially cost-free, but also because part programs will not be specific to any one machine. Machine tools will have sufficient embedded "intelligence" to determine their own parameters for a given part. CIM will ensure that the requisite information is transferred to individual machines from the original CAD data for that part.

This scenario of a CIM and CIM system producing parts in small batches along an optimal processing path (but using multiple processing paths when necessary) for data generated by the CAD system is the basis of flexible automation. The operation of the system will be predicated on additional attributes. To illustrate, the feedback provided by the CAD system to improve the producibility of a given design depends on the production capabilities of the machine tools in the system. Although the system will automatically maintain data on current production capabilities, compromises will undoubtedly arise between maximizing the producibility of a design with current process capabilities and maximizing the functionality of the part or increasing producibility with alternate production processes. The manufacturer may need to compromise with the purchaser to ease the specifications for producibility or increase the price to compensate for a less producible design; subcontract production of the part to a firm with the necessary production capabilities; or add new capabilities to his own production facilities. Even a sophisticated CIM will function effectively only within relative narrow ranges of materials inputs, processing capabilities, and product dimensions. Its potential cost advantages are likely to be threatened when these determinants of production requirements are still subject to substantial change.

The process flexibility discussed at the outset involves these considerations, along with factors such as maintaining optimum equipment use rates while maximizing the overall flow of parts on the factory floor and minimizing routing costs. This type of flexibility also would include the ability to accept data from subcontractors' CIM systems, determine the most efficient producer of the part, and integrate that subcontractor's production and schedule with the other aspects of the total order. If the part is not subcontracted, the CIM system will need to be sufficiently flexible to allow rapid acceptance and integration of new process capabilities. Continued development of new technologies is likely to increase the importance of innovation flexibility; to maintain competitiveness, the CIM system will need to be modular, using computer architectures and interfaces that permit additions to the system to be made easily.

The system should not be so integrated, however, that a single failure can stop the entire factory. Program flexibility should provide every CIM system with sufficient back-up to maintain production even if relatively crucial parts of the system fail. The risk of failure of each component of the system will be weighed against the cost of providing back-up for the component or some other acceptable alternative to maintain production. A wide variety of solutions to this difficult dilemma can be expected.

Effective implementation of these technologies will require adjustments from managers, engineers, and customers. The trade-offs between cost and flexibility will vary among industries and products. Advanced CIM systems will not be infinitely flexible because flexibility will depend on management practices and organizational effectiveness, as well as software, tooling, and material availability. In general, however, both mass production and batch manufacturing industries that can take advantage of CIM technologies can expect a degree of flexibility unknown in the past, with benefits in responsiveness, competitiveness, and total production costs that outweigh the cost of the technology itself.

Cost and Quality

Advanced manufacturing technologies will give managers new tools to help them minimize use of total resources and thereby reduce product life cycle costs. Whether competitive strategy emphasizes low price or product differentiation, price competition in the future is likely to be severe. Reducing life cycle costs and maximizing quality for every product line will be an important determinant of competitiveness and profitability. Cost minimization must not be pursued, however, at the expense of responsiveness and flexibility, as many manufacturers may be tempted to do. The best way to avoid overemphasis on costs is to think in terms of minimizing use of total resources, not only in production but also in purchasing, design, distribution, finance, marketing, and service. Nevertheless, the attention given to individual production factors will continue to depend to a great extent on relative factor costs and the shifting importance of factors in particular industries.

Because new manufacturing technologies will be developed and implemented at various rates, the effect of technology on relative factor costs is difficult to predict. For some manufacturers, new technology is likely to have only a limited effect on direct labor costs and, indeed, will be applied for reasons other than labor savings. The use of CAD and CIM systems will allow these manufacturers to continue to design and produce parts in the United States, but other operations, especially some assembly operations in which labor remains a high proportion of cost, may be candidates for subcontracting or production in low wage countries. The potential savings from low labor rates abroad would need to be balanced against the costs of coordinating demand, production, and delivery. Timely production and delivery will be important in avoiding loss of orders and inventory costs that may not be faced by competitors. These factors will require manufacturers with offshore facilities to use significant forward planning to align production with demand. Advanced technologies will allow manufacturers to handle data in ways that should help to ameliorate the disadvantages of offshore operations, but these gains may not be sufficient to offset the transportation costs, delays, and relative isolation entailed by distant production facilities.

For many manufacturers, advanced technology can be expected to allow more rapid reduction of direct labor inputs, although again, labor savings may not be the major motivation for the investments. A CIM system with flexible automation will permit managers to refine and adjust operations to reflect changes in relative costs over time. It also will introduce entirely new elements to the manufacturer's cost structure, alter traditional ways of measuring costs, and eliminate some major portions of traditional factory costs. It will be possible, for example, to reduce direct labor to insignificant levels or eliminate it in some applications. With no direct labor inputs, some measurements of labor productivity and cost allocation based on direct labor will be irrelevant. New cost accounting systems will be a major need.

Elimination of direct labor is not the same as eliminating labor costs. Technicians, engineers, and programmers will be needed in increasing numbers to maintain and implement CIM systems. Salaries for these employees are likely to exceed wages for direct labor, and their productivity may be more difficult to measure. Even with higher individual salaries, however, labor costs should decline as a share of total production costs because of the capital investments required to keep a CIM system up to date.

It is difficult to predict the effects of investments in new technology on capital costs as a proportion of total costs. Firms will need to monitor the production capabilities of their competitors, as well as those demanded by the market; timely updates of the design and production system will be a competitive necessity. Greater return can be expected and less total capacity may be needed, however, because CIM is expected to allow more work shifts and more optimal use of productive equipment through flexible process plans, less scrap and rework, higher quality production, and lower product life cycle costs. Justification and amortization of technology purchases will be based on total system performance, which implies a significant shift in the measurement and allocation of capital costs.

Although total capital costs in the future are unpredictable, one element of capital costs—tooling costs—can be expected to increase dramatically over historic levels. Continued implementation of CIM systems, along with the emergence of new materials and processes, will greatly increase the volume, variety, and complexity of tooling requirements and the need to move the tooling around the factory. Increased tooling and handling costs are already evident in many manufacturing operations, and the trend can be expected to accelerate as other new developments are implemented.

Developments in new materials and material processing will have a significant impact on material costs and availability, especially vis-à-vis product performance and quality. Advanced material handling systems should have a major effect on the costs of moving and storing materials. New materials such as high temperature ceramic superconductors, high strength resins, composites, and ceramics will create new options in product development, providing significant improvements in performance while reducing material requirements. Ceramic engine parts, for example, are under development by virtually all major combustion engine manufacturers and will allow simplified engine design and fewer total parts. Once the material and processing problems are overcome, the effects on material costs and requirements will be substantial. Similar effects can be expected with other materials and applications. Even with more traditional materials (for example, metals), progress in ultraprecision machining will reduce material requirements and improve product performance. These developments will greatly expand the choices available to managers in material application and processing, which will make the data management abilities of CIM virtually indispensable.

For all manufacturers, the ability to accumulate, store, and process data will become a growing force in production, motivated by a rapid decline in the cost of data management. Data will be gathered and accessed rapidly and easily, with a number of important repercussions, one of them being the increasing significance of time as a factor of production. The time between design and production and from order to delivery will shrink dramatically. The trend toward shorter product life cycles and rapid technological developments can be expected to make very small increments—hours and days instead of weeks, months, or even years—crucial factors for competitive production. The increased ability to manipulate and accumulate all types of data will have a significant impact on plant location decisions. Since information for the entire organization will be available almost concurrently regardless of plant site, the criteria for plant locations will emphasize costs, available process technology, responsiveness, quality, and optimal resource use. Some of the considerations involved in decisions to move plants offshore have been discussed; these considerations may lead to more decisions to keep manufacturing capacity onshore.

In fact, there is some evidence that, due to responsiveness, flexibility, and quality concerns, future trends in factory locations, particularly for component manufacturers, will be toward a proliferation of smaller factories closer to final markets and greater use of contiguous manufacturing, in which progressive manufacturing operations are located in close proximity to each other. New technologies will make both of these strategies easier to pursue for many industries, and market demands may make them a necessity. For some industries, the concept of the microfactory will become important: small factories, highly automated and with a specialized, narrow product focus, would be built near major markets for quick response to changing demand. Because of the unique circumstances of each industry, in terms of technology availability, labor requirements, cost structures, and competitive circumstances, it is difficult to predict how strong each of these trends will be, but they are representative of new options available to manufacturers in their efforts to maximize competitiveness.

In addition to the costs of labor, capital, materials, and data management, CIM can be expected to change other traditional costs in manufacturing. The producibility feedback and modeling capabilities of the CAD system will reduce product development costs, which will allow firms to either reduce prices or do more product development. The monitoring and self-correcting mechanisms embedded in the machine tools will reduce scrap and rework costs. These capabilities also will result in higher quality, which will greatly reduce service costs and attract a broader customer base. Inventory costs also will decline because set-up times and costs will not be major considerations in the flexible CIM system, it should therefore be possible to make very small batches of products to order without large inventories of materials or finished parts. Finally, the flexibility of CIM will permit companies to use materials with a variety of specifications, customize products, and focus on markets for products with higher value added. The extent to which companies pursue these capabilities will depend on their overall competitive strategies.

All of these considerations imply that manufacturers will have a very different cost structure in the future than they have today. Continuous investment will be a competitive necessity and will be justified on a systems basis. The CIM system will manage the use of material, time, and equipment to such an extent that total inputs and, therefore, total production costs, can be optimized within the limits imposed by the hardware and software capabilities of the moment. Close monitoring of input requirements will be a crucial ingredient, along with responsiveness and flexibility, in determining competitiveness. High quality will be a necessity because customers will expect it; any perceived slippage in quality will cost customer loyalty and market share and obviate many of the benefits of CIM and new materials.

Conclusion

The most important factor in improving responsiveness, flexibility, costs, and quality will be the effectiveness of management practices, organizational design, and decision-making criteria. As the capabilities and advantages of new manufacturing technologies progress, they will become increasingly important to managers' future strategies for improving competitiveness. Furthermore, the effectiveness of the technology in accomplishing corporate goals is likely to depend on having the most appropriate hardware and software in the CIM system. Changes and adjustments to the system will be based on each company's market situation, product line, and customer base, so many of these capabilities will be internally developed and proprietary. Along with the management aspects of the manufacturing organization, they will determine competitive advantage in the manufacturing environment of the future.

This view of manufacturing technology is very different from the traditional technical view. Advanced manufacturing technologies are not going to solve all the problems of production. Instead, they will give managers many more options. Managers will have an even greater need to focus the goals of the firm and then assess the needs of the manufacturing function and how technology can best address them. Once choices are made, managers will not have the luxury of running the technology for long periods while they focus on product design, marketing, or some other function to maintain a competitive position. Dynamic, continuous improvement of manufacturing capabilities will become essential to long-term success.

People and Organization

The competitive pressures and technological capabilities discussed in this book are two dimensions of the changes that can be expected in the future manufacturing environment. This chapter addresses changes in the management of people and organizational design that future manufacturers will need to pursue to be successful. Such changes can strengthen the competitiveness of many companies regardless of the technology employed, and in virtually every case, modifications in both the internal and external relationships of the business are a prerequisite to effective use of new technology.

The changes needed in people and organizations will be a difficult aspect of the revolution in manufacturing. They require a dramatic refocus of the traditional culture in the factory, away from hierarchical, adversarial relations and toward cooperative sharing of responsibilities.

With such fundamental changes, progress will be slow, the degree of change will vary among companies, and the full transition is likely to be accomplished by a relatively small number of companies. However, the demands placed on manufacturers to be effective in an increasingly competitive marketplace can be expected to push managers and workers in the directions described in this chapter.

Much depends on the size and culture of the firm and the commitment of managers and workers. Many manufacturing enterprises need changes not only in broad organizational areas and management philosophy but also in employee behavior, union policies, and customer-supplier relations. Every stakeholder—managers, employees, owners, suppliers, and customers—must recognize the challenge and be prepared to change traditional practices. Furthermore, people who may not have a direct stake in manufacturing—government officials, educators, researchers and scientists, and the general public—will need to understand the importance of manufacturing to future prosperity, recognize the evolving role of manufacturing in the U.S. economy, and support the many social and cultural changes that will both result from and encourage continued progress in U.S. manufacturing.

A Systems Approach

Part of the problem of U.S. manufacturing is that the common definition of it has been too narrow. Manufacturing is not limited to the material transformations performed in the factory. It is a system encompassing design, engineering, purchasing, quality control, marketing, and customer service as well as material transformation; the operations of subcontractors and the whims of customers are also important parts of the system. The systems approach is a key principle not only for manufacturing technology, but also for organizational structure, supplier relations, and human resource management. Such a concept has been foreign to most U.S. managers (although embraced by Japanese managers), and the result has been a lack of responsiveness and declining competitiveness in many industries. Managing manufacturing as a unified system will profoundly affect every activity involved; it is the only way to take advantage of the many opportunities in both products and processes that the future will bring.

An aggressive systems approach in a company should eliminate many of the functional distinctions that can introduce inefficiencies into the production process. Instead of the labyrinth of functional departments common in many firms, the operations function is likely to become the focus. Ancillary and supportive functions will be reintegrated into operations. Maintenance and process design, for example, will no longer be distinct entities with separate schedules and staffs; instead, employees in operations will be responsible for maintaining equipment or modifying the process as the need arises. Such reintegration will mean that management structures are likely to be more streamlined and that many job classifications will be eliminated to allow employees to perform multiple tasks. Job design and classification will be based on broad operational functions rather than narrowly defined activities.

Functions such as product design, manufacturing, purchasing, marketing, accounting, and distribution will require close cooperation and tight coordination. Eliminating them as separate departments would be impractical—the various types of expertise will still be needed—but, with increasing computerization and communication capabilities, information on each area will be widely available and close cooperation will be essential. This cooperation often is likely to be accomplished through working groups of people from different departments. They will include permanent groups to ensure long-term integration of ideas and temporary groups designed to address specific projects. Techniques such as comprehensive job rotation may be used to eliminate interdepartmental barriers. The process of integrating the data bases and process technologies in the factory also will help to eliminate artificial barriers between functions, but the major tools for change will be the guidance of senior managers and the initiatives of employees.

In external relations, a systems concept calls for reassessment of the mechanisms used to specify, order, manufacture, and deliver subcontracted parts. Because production by suppliers will be viewed as the initial step in the manufacturing system, major customers will need to take a strong interest in the capabilities of their suppliers and institute programs to raise those capabilities through gentle persuasion, direct assistance, or reselection. As an example of the changes that can be expected, customers' design equipment will be able to communicate directly with suppliers' production equipment. Substantial investments will be made in communication linkages to allow extensive sharing of data on design, production scheduling, material requirements planning, and costs.

These types of arrangements will be essential for flexible management of the manufacturing system, but they imply significant change in supplier relations. More subcontracts will be long term, and the number of captive shops supplying one customer can be expected to increase substantially. The investment in communication links by the customer and the corresponding investment that the customer will expect of the supplier will make long-term contracting desirable for both parties. Since long-term contracts weaken the threat of changing suppliers if standards are not met, a strong commitment to close cooperation will be a necessity. Problems will need to be solved as they arise, just as with in-house production, because the cost of failed relationships will be high. Both parties will lose independence in the subcontracting process, but the advantages of an integrated, highly efficient manufacturing system will outweigh the costs.

Examples of these types of relationships can be seen already in both the automobile and aircraft industries. Long-term subcontracting has been common in both industries for years, but only relatively recently have design data been transferred directly from the CAD system of a major manufacturer to the machine tools of the subcontractor.

Participation and Ownership

A key step in the evolution of human resource management in manufacturing will be to broaden participation in the company's decision-making process. Employees at all levels should be given an opportunity to contribute ideas, make decisions, and implement them in areas that may affect operations beyond the individual's formal responsibilities. The principle involved is intellectual ownership: If all employees can feel a degree of ownership in decisions that affect them and the company, they are likely to support those decisions more enthusiastically, resulting in a highly motivated work force and a more responsive, effective company. Extensive, even universal, participation in decision making gives all employees a stake in the company, beyond financial considerations, that may be essential for continued competitiveness in a rapidly changing environment.

For most manufacturers, such a decentralized decision-making process will require a major cultural shift and a number of prerequisites to avoid disorder. The most fundamental requirement is a well-understood, common set of goals and a high level of commitment to them from both managers and employees. Beyond that, both management and labor must meet certain responsibilities.

Management cannot expect employees to contribute ideas and participate in decision making without the necessary knowledge and expertise. Vehicles will be needed to facilitate the rapid flow of information within the organization and to ensure that the proper intellectual resources are available at all levels. Close links between upper management and operatives on the factory floor will be required for rapid information exchange and responsiveness. Information must flow both upward and downward in the organization. Employees must fully understand the goals and priorities of the firm to make consistent, effective decisions.

Managers need to be assured that the correct decisions are, indeed, being made. This type of cooperative, two-way flow represents a radical shift for many firms that may cause significant cultural disruptions.

Information linkages are currently provided by several layers of middle management that serve primarily as an information conduit. With decision-making responsibility pushed to the lowest possible level, the extra layers of middle management are likely to be both unnecessary and unaffordable. The result in most cases probably will not be mass layoffs of middle managers; instead, the change will become manifest as a gradual blurring of the distinctions between operatives and managers. Middle managers will be reduced in number and merged into new roles that allow direct access between upper management and floor workers. Knowledge requirements, authority, and responsibility will tend to converge, resulting in much flatter organizations. This fundamental change in organizational structure already is happening in a number of companies.

Progress in factory communications technology also can be expected to facilitate information flows and contribute to the elimination of management layers. Wide, if not universal, access to all types of information, from part designs and scheduling to accounting data and marketing plans, will reduce the need for personal exchanges of information. Both upper managers and operatives will have direct access to the information needed to make effective decisions. With the process of flattening the organizational structure and creating direct communication channels between operatives and upper management well underway, advanced data tracking and communications systems will be far more effective. Without the efficiencies introduced by wide participation and shared responsibility, a manufacturer might be overwhelmed by the volume of information provided by the new technologies. As with most aspects of the new manufacturing environment, organizational and technological changes complement each other and cannot be separated without tremendous costs in corporate effectiveness.

Effective decision making depends on trained personnel. Employees not only must be competent in their own jobs, but also must understand the relationship of their jobs to overall corporate operations and goals. Developing the required knowledge will require extensive training, by a variety of mechanisms and a significant amount of job rotation at every level of employment. As new technology is implemented, substantial training can be expected from equipment vendors, but the broader scope of job responsibilities is likely to require off-site classroom courses, companywide seminars, and on-the-job training. Training will need to be extended to every employee as a corporate necessity, and it will account for a much higher percentage of working time and total costs than it has traditionally. Job rotation also can be expected to be far more extensive than has been common in U.S. firms. Short-term efficiency will be lost to a degree as employees rotate into unfamiliar jobs, but the long-term benefits, in terms of systems knowledge and a greater sense of a corporate team, will be indispensable for competitive operations.

Employment Security

Creating a companywide environment suited to competitive production with advanced technology will require a strong commitment to employment security: The ability of a firm to retain its employees even though changing market conditions or advancing technology may significantly change the content of the jobs they do. There is little doubt that manufacturing will no longer be a strong source of employment for the unskilled and semiskilled. Factories will employ fewer people in these groups in particular, but the number of people employed among skilled workers and managers will also decline. The remaining jobs, however, will be challenging, rewarding, and in demand. Competition for those jobs and competition for good people will make strong job security an interest of both employee and employer. Information systems, training programs, and changes in organizational structures will represent huge investments in human capital. Cyclical layoffs or unnecessary turnover would severely limit the return on that investment and risk a complete breakdown in the company's operations. Employment security is crucial to engendering commitment of workers as true stakeholders. Increased responsibility and participation will improve the attractiveness of manufacturing jobs, but employees are not likely to feel a strong stake in the company unless they believe it has a stake in them. From the perspective of both company and employees (unionized or not), job security is a critical principle to pursue, which in itself will represent a significant change in the attitude of management and labor.

Although various mechanisms will be used in pursuit of a stable work force, absolute job security is likely to remain both elusive and a source of contention. Some companies recently have been very successful in relocating unskilled and semiskilled workers to other plants and in retraining them to perform new and varied tasks in the automated factory. Many employees, however, will be unable to adapt to a new environment that requires more skills, knowledge, and responsibility. Following this shakeout, however, it will be feasible and advantageous for employers to provide strong job security for a core group of employees. This core group would be capable of handling daily operations, and subcontracted temporary workers would be hired to meet surge demands.

These temporary crews would perform specific duties that do not require extensive knowledge of the company's operations. They would be managed by the operatives usually responsible for those tasks to maintain continuity in decision making. This approach will insulate the core staff from fluctuations in market demand (essentially making the core labor a fixed cost), will provide employment opportunities for previously displaced workers, and, if widely used, may change the nature of unemployment trends in the macroeconomy. The approach is already used extensively in the airframe industry.

Job security also may be strengthened by the trend to perform previously subcontracted work in-house, although this trend will vary across industries and firms depending on size, available technology, and product mix. No company can do everything well, and, in many industries, subcontracting is a way to share risks, costs, and expertise. In some industries, however, advanced process technologies are likely to provide sufficient capacity and flexibility to encourage firms to pull subcontracted production in-house. The advantages in quality control, production scheduling control, and design change would reinforce the job security benefits of such a strategy. In fact, the trend can be seen already in the domestic mainframe computer industry. In other industries, the advantages of small, focused factories may create more subcontracting than has been traditional. Despite these variations, many companies will find that the advantages of in-house production outweigh the disadvantages, particularly in maximizing return on the large investment in human capital.

Incentives, Evaluations, and Decision Criteria

Traditional measures of success in manufacturing will be inadequate for tomorrow's manufacturing environment. New measurements, as well as new rewards, will be needed to manage production effectively, maintain employees' motivation, justify new investments, and stimulate stockholders' interest.

In the area of factory operations, managers will need new criteria on which to base operational decisions. Mechanisms for improving factory effectiveness, such as precise inventory control systems, material requirements planning, in-house production of previously subcontracted work, production process planning, and the many aspects of factory automation will change the operations of the future factory. The criteria that managers have traditionally used to make operational decisions will change, and, in many cases, the decisions will change. A factory using just-in-time inventory control, for example, will have less input inventory on hand than a manager may have been accustomed to having. As another example, changes in production processes and work flows will change the criteria used to judge effective machine utilization rates, manning levels, acceptable work-in-process inventory, and tooling requirements. Managers will need retraining to alter their thinking about the effective operation of the factory to prevent old habits from inhibiting potential cost savings, quality improvements, and overall effectiveness.

Evaluations of individuals—both managers and workers—are likely to be much more subjective than they have been traditionally. Quantifiable improvements in individual performance, such as increasing output per hour or shift, will not be applicable to automated, integrated production with emphasis on project teams. Objective indicators of performance will remain in areas such as quality, delivery, process system costs, customer satisfaction, and company earnings, but these will reflect more on group efforts or the total work force than on individuals. Consequently, individuals' pay is expected to shift from hourly wages to salaries; pay will entail a greater emphasis on bonuses based on improvement in short-term results, long-term improvements in the total system, and achievement of the goals of that particular level of the organization. The "profit center" and "cost center" focus used in the past as a basis for judging individual performance can be expected to be replaced by a systems focus. Subjective assessments of individuals' skills and competence by their peers will affect salary decisions at least indirectly. Promotions will remain a form of recognition and opportunity for increased responsibility, but increased use of project teams and job rotation is expected to diminish the importance and obvious benefits of promotions; the elimination of most middle management positions will reinforce this trend.

At the company level, evaluation of performance will depend largely on a meaningful management accounting system. Traditional methods that aggregate data, allocate costs based on direct labor, and compute data over long intervals (usually monthly) will be ineffective and counterproductive in the new environment. With computer-integrated manufacturing technologies, companies will be able to measure their performance continually by resource category, cost center, and product class and will abandon cost accounting systems based on direct labor. New accounting systems will give manufacturers the accurate, timely data they need to respond rapidly to changing conditions. The availability of more relevant data will give almost everyone in the business a clear perspective on total performance and its response to key decisions on matters such as investment, personnel, subcontracting, and research and development expenditures.

Changes in accounting procedures will contribute to the strong trend toward balancing short-term results against long-term prospects in determining the health of a manufacturing firm. New criteria will be developed to give stockholders and investors a basis for assessing the steps a firm is taking to ensure its long-term competitiveness. These criteria may include research and development expenditures, the amount and kinds of investment over a given period, training and recruitment patterns, and the activities of major competitors. None of these indicators will be conclusive, but as a package they will give stockholders more information than is common currently.

Future Focus

In the new manufacturing environment, efficiency alone will not ensure success. Foresight will be the ultimate competitive weapon, because market share and profit margins are likely to be small for the followers. A long-term, future-oriented focus, extending beyond the next quarter or year, will be a competitive necessity. Manufacturers will need to devote an increasing amount of time, money, and energy to those parts of the business that will have a preponderant impact in the future, particularly product and process research and development.

The pace of change is expected to be rapid, so emphasis on strong in-house research and development will be a necessity for firms seeking a leadership position. Manufacturers will need to accept the risks inherent in long-range research. Investments in scientific and engineering personnel, laboratories, and computers are expected to be a significant portion of total capital budgets. At the same time, the need to implement new technologies, introduce new products, and attract talented personnel will be expensive. Some companies will share costs by participating in research consortia, which in fact may be the only viable method of research in some industries. Other companies will save research costs by using licensing agreements, but as product life cycles shorten, the value of licensing as a relatively inexpensive way to enter new markets can be expected to diminish. Companies will face difficult choices in striking a balance between spending for future and immediate competitiveness. Similar circumstances exist today. The major difference is that future manufacturers will probably have much less ability to milk profits from new products because competing entrants will be close behind. The costs of being a follower will be more apparent, so the weight given to future-oriented investments should be much greater than it has been traditionally.

Conclusion

With these changes in the human and organizational components of manufacturing, the factory will become a much different factor in society. Although opportunities for unskilled or semiskilled labor will diminish, the jobs that will be created are expected to be challenging and of high quality. Also, manufacturing jobs will be in demand among graduate engineers, who do not generally prize them today, and there may be too few to go around. Firms will have such large investments in people that they will make extraordinary efforts to retain employees, which will limit job creation at existing plants. This constraint may be countered somewhat by the trend in some industries toward microfactories, although the labor requirements of such facilities may be quite small. Employment opportunities also will arise in industries producing goods yet to be invented and in the variety of services that can be expected to develop to support future manufacturers.

These changes in the factory will permeate the social and economic fabric of the nation. Changes in internal factory operations will affect relations with unions, subcontractors, wholesalers and retailers, producers of services, and other economic activities outside but closely related to factory operations. The expectations and opportunities of workers at all levels will be affected by the cultural revolution that has already begun in manufacturing. For many companies, the rate and direction of change will be determined through the collective bargaining process; for other firms, less formal approaches of labor-management cooperation will be used. None of the changes will be sudden, however, and no two industries will progress at the same pace.

In fact, there may very well be a backlash from both managers and workers who have a strong stake in traditional relationships and organizational structures. Consequently, the changes in people and organizations will, at best, proceed in fits and starts, but the benefits in terms of manufacturing effectiveness and profitability are expected to be so clear that these difficult cultural changes will be implemented. The specifics of these changes are difficult to predict because they are based on individual decisions in a vast variety of circumstances. The direction of change, however, is becoming increasingly clear, and the repercussions will be wide-ranging.

Plan for Tomorrow

Manufacturing has already entered the early stages of revolutionary change caused by the convergence of three powerful forces:

• The rapid spread of manufacturing capabilities worldwide has created intense competition on a global scale.
• The emergence of advanced manufacturing technologies is dramatically changing both the products and processes of modern manufacturing.

• There is growing evidence that changes in traditional management and labor practices, organizational structures, and decision-making operations, provide new sources of competitiveness, and introduce new strategic opportunities.

The effects of these forces are already being felt by the U.S. manufacturing community. Domestic markets that were once secure have been challenged by a growing number of foreign competitors producing high quality goods at low prices. New technologies are helping U.S. manufacturers compete, but many technical and social barriers remain before advanced technologies have a major, widespread impact on manufacturing operations. Unfortunately, foreign competitors may well have overcome some of these barriers first, using new technologies to increase their competitiveness.

As these points indicate, the three trends now affecting manufacturing are closely interrelated. Increased competition has demonstrated the need for U.S. manufacturers to reexamine traditional human resource practices and their use of new product and process technologies. Corrective measures, however, cannot focus exclusively on either area, since technology will not be effective without changes in human resource practices, and the benefits from those changes are limited without the productive thrust offered by new technologies. Meanwhile, the competition intensifies, current production must be maintained, and the resources available to make the required changes always seem inadequate.

All of this poses a difficult dilemma for manufacturers who have depended on stability to maintain competitive production. Many manufacturers recognize the need to adapt, but do not know what changes are necessary or how to implement them. More than anything else, the key problem is that the forces affecting manufacturing require that managers think and act differently to bring about change in a systems contest and that workers accept new roles and responsibilites.

The major roadblocks to more competitive U.S. manufacturing are in the attitudes, practices, decision-making criteria, and relationships of both managers and workers. Chapter 3 describes the kinds of practices that are likely to be required for future manufacturing competitiveness. That vision means that hierarchical, adversarial management structures will handicap attempts to improve competitiveness. Employees at all levels of the organization will need to be viewed as a resource, and the organization will need to be structured so that everyone will have the opportunity and responsibility to make the maximum contribution. Furthermore, the importance of the manufacturing function in the total corporate context will need to be recognized. Functional integration based on a clear understanding of the manufacturing systems concept will be a major key to competitive success. This way of thinking about manufacturing is foreign to most managers, workers, and educators in this country, and it may be overly optimistic to expect such a dramatic shift in attitudes and culture. Ingrained attitudes will be difficult to change and may require a generational shift.

This book provides some direction, not a solution. Circumstances vary too much to try to prescribe specific actions, but the direction for change should be clear. The use of new advanced manufacturing technologies is insufficient. The key is to focus on evaluating traditional managerial practices, relationships, decision-making criteria, and organizational structures to determine specific strengths in responding to competitive pressures. The renewed organization will be in a better position to implement new technologies and further strengthen competitiveness. For some companies, however, attempts to implement new technologies will force labor and management changes. Managers will need to realize that implementation of advanced manufacturing technologies to automate existing processes will yield suboptimal results. Efforts to optimize the technologies will demand creative thinking to take advantage of the opportunity to redesign many processes, simplify many designs, and change the flow of work on the factory floor. This creative thinking and the necessary cultural changes will be the major obstacles to attaining improved competitiveness.

Government should play a significant role in encouraging and supporting the changes in manufacturing, but the impetus must come from private companies. In general, the main responsibility for government is threefold: (1) to recognize the importance of a strong manufacturing sector as a source of goods for international trade and as a crucial factor in continued economic prosperity and strong defense; (2) to support the process of change in manufacturing; and (3) to stay abreast of the changes taking place in manufacturing and adapt government policies and programs to maximize their effectiveness in the new environment. In addition, some specific government activities, for instance in education, research, and tax incentive programs will need to be particularly sensitive to manufacturing requirements and ensure that necessary resources remain available.

In summary, U.S. manufacturers are facing a crucial challenge. Traditional markets are being attacked by imports and traditional practices are not producing adequate results. Changes in labor and management attitudes, organizational design, and the role of the manufacturing function in the total corporate system are needed to regain and maintain competitiveness. New technologies will help this process, but manufacturing strategies will need to be evaluated to ensure both that the right technologies are used and that the full potential of those technologies is realized.

Chapter 2

Description of CIM
System Elements

Introduction to Computer-Integrated
Manufacturing Systems

This chapter will provide both a description of CIM systems and discussions of the important problems and directions for development of the sub-element technologies. Included are computer-aided design (CAD) and computer-aided design and drafting (CADD) and its interface with computer-aided manufacturing (CAM) systems.

The essential difference between conventional factory machines and the automated production inherent in CIM is the latter's integration of all information technology required to design, produce, and deliver the product. The use of computers and communications systems allows machines to perform a greater variety of tasks than fixed automation allows, while also automating some tasks previously requiring direct human control.

Computer-aided design can address the central problems of manufacturing. These include enhancing information flow, improving coordination, and increasing flexibility and efficiency (defined respectively as the range of products and the volume of a specific product which a factory can produce economically). Manufacturers can increase their productivity and control over the manufacturing process.

Though labor savings are the most obvious benefit of automation, savings through more efficient use of materials may be more significant in many manufacturing environments. Computer-integrated manufacturing systems can reduce waste, reduce levels of finished product inventory, and reduce the manufacturer's substantial investment in the products that are in various stages of completion, known as "work in process."

Some technical factors which hold back CAD's potential uses in manufacturing include the lack of standardized programming languages, and the traditional (nonmechatronic) organizational barriers in industry (for example, between design engineers, production engineering, and manufacturing). Nevertheless, the technology appears to be quite adequate for the vast majority of near-term applications, including a significant backlog of available tools which manufacturers have only begun to exploit.

The use of computer-aided design tools in computer-integrated manufacturing systems is much more powerful than their use for single tasks or processes. Such integration not only magnifies the productivity and efficiency benefits of computer-integrated manufacturing systems but also induces changes in all parts of the factory including management strategies, product designs, and material flow.

Many industrialists have a futuristic vision of CIM that includes maximum use of coordination between CAD/CAM tools, with few if any human workers, while others downplay CIM as a revolutionary change and emphasize that factories will only adopt automation as they deem appropriate. Most agree that the widespread use of CIM and virtually unmanned factories is likely to arise around the turn of the century.

Work on the future development of CAD technologies emphasizes increasing their versatility and power, enhancing their capability to operate without human intervention, and developing integrated tools. Researchers and industry spokesmen report progress in virtually all these areas, but the problems in CIM are sufficient to keep researchers busy for many years to come. According to many experts, the 1990s may bring many major technical advances which could significantly expand the applications for CAD.

The recent technological and economic trends, including improvements in computer control, improvements in equipment interfacing, cost reductions, international competition, and a growing interest in manufacturing productivity, have fueled rapid growth during the last few years (Figure 2–1). Although automation industries are growing rapidly, much of their impact on the economy will be realized indirectly. This is because their principal customers are other businesses, which adopt automation to use in producing consumer and other producer goods, from appliances to construction equipment. The broader the customer base for CIM systems, the greater the direct economic contribution of the automation businesses will be seen. For example, the machine tool industry, a principal supplier of capital goods to metalworking manufacturing industries, is small in terms of output and employment (under 70,000 employees in 1983 and under 80,000 employees in 1982, down from about 100,000 in 1980; about two-thirds are production workers). By contrast, the computing equipment industry, which is less labor-intensive than the machine tool industry and which serves both industrial and consumer markets, is much larger (employing about 3,420,000 in 1982).

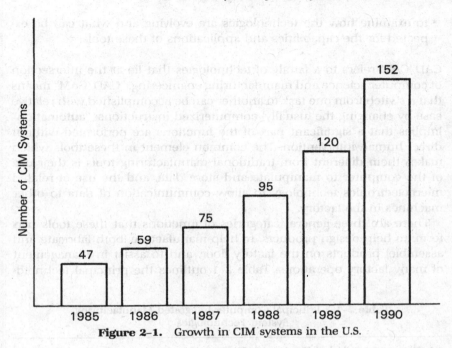

Figure 2-1. Growth in CIM systems in the U.S.

At present, U.S. producers dominate U.S. markets for CAD/CAM. They also export automation products, and some U.S. firms have invested in the production of CAD/CAM equipment and systems abroad. For example, Westinghouse/Unimation has a robot plant in Telford, England, and Cincinnati Milacron has several European machine tool plants.

Near-term growth of domestic CAD industries will continue to depend on favorable economic conditions. Industry analysts forecast rapid CAD market growth. For example, Predicasts, Inc., has forecast that the combined market for "manufacturing computers, CAD systems, machine tools and controls, and robots will grow over 15 percent annually." Sales will double, according to that analysis, to almost $15 billion by 1987. The analysis assumes a GNP growth rate in real terms of 3.8 percent per year, and a healthy U.S. automobile industry. Industry growth will also depend on the ability of American managers to justify investments in CAD and to become adept at using it.

The purpose of this chapter is:

• to describe the CIM systems that together comprise computer-aided design (CAD) and computer-aided design and drafting (CADD) and its interface with computer-aided manufacturing (CAM) Systems,
• to evaluate their usefulness for manufacturing,

- to examine how the technologies are evolving and what can be expected for the capabilities and applications of these tools.

CAD/CAM refers to a family of technologies that lie at the intersection of computer science and manufacturing engineering. "CAD/CAM" means that a switch from one task to another can be accomplished with relative ease by changing the (usually) computerized instructions; "automation" implies that a significant part of the functions are performed without direct human intervention. The common element in these tools which makes them different from traditional manufacturing tools is their use of the computer to manipulate and store data, and the use of related microelectronics technology to allow communication of data to other machines in the factory.

There are three general categories of functions that these tools perform: to help design products, to help manufacture (both fabricate and assemble) products on the factory floor, and to assist in management of many factory operations. Table 2–1 outlines the principal technolo-

Table 2–1.   Principal Computer-Integrated Manufacturing
System Technologies

I.   Computer-Aided Design (CAD)

  A.   Computer-Aided Design (CAD)

  B.   Computer-Aided Drafting (CADD)

  C.   Computer-Aided Engineering (CAE)

II.   Computer-Aided Manufacturing (CAM)

  A.   Robots

  B.   Machine Vision

  C.   Numerically Controlled (NC) Machine Tools

  D.   Flexible Manufacturing Systems (FMS)

  E.   Automated Materials Handling (AMH) and

    Automated Storage and Retrieval Systems (AS/RS)

III.   Tools and Strategies for Manufacturing Management

  A.   Data-Driven Management Information Systems (DDMIS)

  B.   Computer-Aided Planning (CAP) and

    Computer-Aided Process Planning (CAPP)

---

gies included in these categories, each of which will be described in this chapter.

The three categories of automation technologies—tools for design, manufacturing, and management—are not mutually exclusive. In fact, the goal of much current research into computer-integrated manufacturing systems is to break down the barriers between them so that design and manufacturing systems are inextricably linked. However, these three categories are useful to frame the discussion, particularly since they correspond to the organization of a typical manufacturing firm.

Discrete Manufacturing

Some background about manufacturing is important to provide a context for assessing the usefulness of CAD tools. CAD can affect many kinds of industry. Discrete manufacturing is the design, manufacture, and assembly of products ranging from bolts to printed circuit boards. Nonmanufacturing applications include architecture, or continuous-process production methods; for example, chemicals, paper, and steel.

Electronics manufacturing industries do not fit neatly into a discrete vs. continuous-process classification. Some areas, particularly the fabrication of semiconductors, most resemble continuous-process manufacturing. Other portions such as printed wiring board assembly are more discrete. Electronics industries have been leaders in both producing and using computerized factory automation.

Discrete manufacturing involves metalworking for mechanical applications: shaping, forming, and finishing metals into usable products such as engine blocks. However, an increasing proportion of mechanical parts manufacturing involves plastics, fiber composites, and new, durable ceramics. These new composite materials both enable new production processes and are themselves affected by automation technologies.

Discrete manufacturing plants can be categorized by the quantity of a given part produced. As Figure 2-2 indicates, discrete manufacturing represents a continuum from piece or custom production of a single part to mass production of many thousands of parts. Mass production accounts for only 20 percent of metalworking parts produced in the United States, while 75 percent are made in "batch" environments. The definition of a "batch" varies according to the complexity of the part and the characteristics of the industry. In batch manufacturing there is not enough volume to justify specialized machines (known as "hard automation") to produce the part automatically. The direct labor involved in fabricating products in batches is a large proportion of the cost of the item. The characteristics of batch manufacturing—its preva-

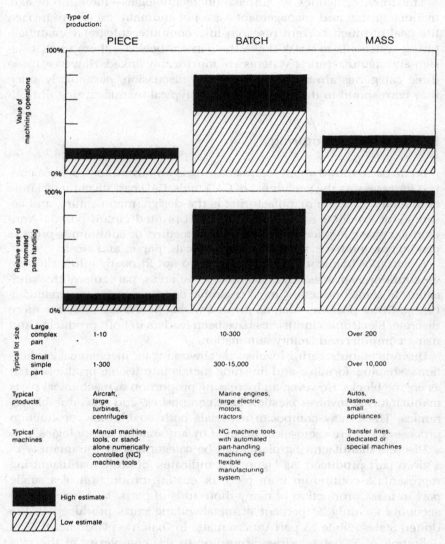

Figure 2–2. Characteristics of Metalworking Production, by lot size.

Source: Office of Technology Assessment

lence, and its low level of automation and correspondingly high level of labor content—are important because they suggest a broad range of uses for computer-integrated systems.

The Manufacturing Process

Figure 2–3 shows the organization of an integrated metalworking manufacturing plant. Most of the elements in this diagram are present in some form in each plant, although factories vary tremendously in size,

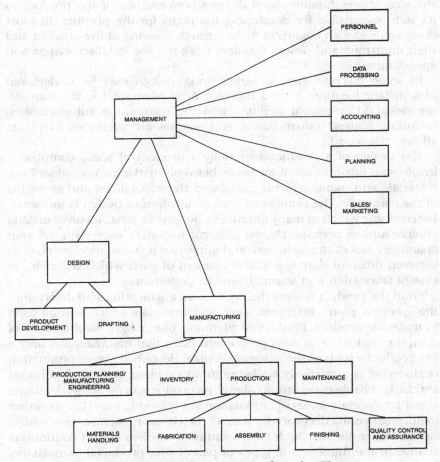

Figure 2–3. Organizational Diagram of a Manufacturing Firm

Source: Office of Technology Assessment.

nature, variety of products, and production technologies. One automobile factory in New Jersey, for example, assembles 1,000 cars per day in two models (sedan and wagon) with 4,000 employees. A small Connecticut machine shop, by contrast, employs 10 people to make hundreds of different metal parts for aircraft and medical equipment, typically in batches of approximately 250 parts.

The manufacturing process usually begins when management decides to make a new product based upon information from its marketing staff, or (in the case of the many factories that produce parts for other companies' products) management receives a contract to produce a certain part (see Figure 2–4). Management sends the specifications for the size, shape, function, and desired performance of the product to its staff responsible for developing the plans for the product. In most companies, design engineers make a rough drawing of the product and then draftsmen and design detailers work out the detailed shapes and specifications.

In some discrete manufacturing firms, design may be undertaken at a distant location, or at a different firm. Automobiles, for example, are designed at central facilities, and the component subassemblies, including bodies, transmissions, and engines are produced in plants all over the world.

The design of a product, especially a product of some complexity, involves an intricate set of trade-offs between marketing considerations, materials and manufacturing costs, and the capabilities and strengths of the company. The number of choices involved in design is immense. Determining which of many alternative designs is "best" involves making choices among perhaps 100,000 different materials, each with different characteristics of strength, cost, and appearance; it also involves choices between different shapes and arrangement of parts which will differ in ease of fabrication and assembly and in performance.

From the product design, the production engineering staff determines the "process plan": machines, staff, and materials which will be used to make the product. Production planning, like design, involves a set of complex choices. In a mass production plant that manufactures only a few products, such as the automobile plant described above, production engineering is a relatively well-structured problem. With high volumes and fairly reliable expectations about the products to be made, decisions about appropriate levels of automation are relatively easy. On the other hand, engineering decisions for a small "batch" manufacturer production can be rather chaotic. Such an environment involves almost continuous change in the number and types of parts being produced (size, shape, finish, material), the tools and levels of skill needed to produce them, and unpredictables such as machine breakdown and inventory control problems.

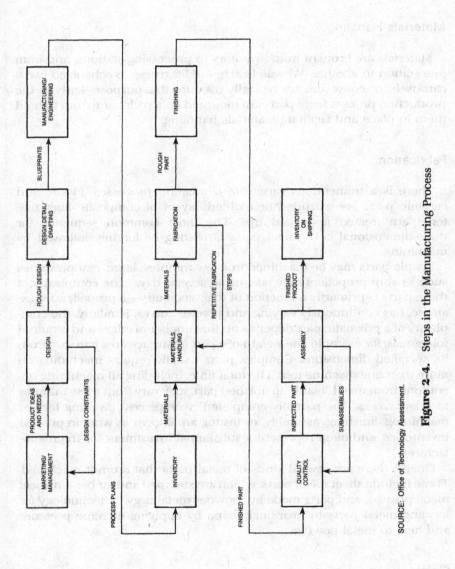

Figure 2-4. Steps in the Manufacturing Process

SOURCE: Office of Technology Assessment.

The steps in production are immensely varied, but most products typically require the following.

Materials Handling.

Materials are brought from inventory to processing stations, and from one station to another. Wheeled carts, forklift trucks, mechanized carts, carousels, or conveyors are typically used for this purpose. Early in the production process large parts are mounted on a pallet or fixture to hold them in place and facilitate materials handling.

Fabrication.

There is a tremendous variety of fabrication processes. Plastic and ceramic parts are extruded or molded; layers of composite fiber material are treated and "laid up." The most common sequence for three-dimensional (3-D) metal parts is casting or forging, followed by machining.

Simple parts may be machined in a few minutes; large, complex ones such as ship propellers may take up to several days. The complexity of these parts is primarily a function of their geometry—a propeller, for example, has continuously varying and precise curves. Similarly, the complexity of a prismatic part depends on the number of edges and required tolerances; for example, the weight of a part or surface area can vary from its specified dimensions. Complex parts usually require machining on more than one machine tool. The total time, including all machining operations from metal "blank" to finished part, may vary from a few minutes to a few weeks. The partially completed "workpieces" awaiting further machining, finishing, assembly, or testing are known as work-in-process inventories, and often represent a substantial investment for the manufacturer.

Finally, there are several kinds of metal parts that are not machined. These include sheet metal parts, which are stamped and/or bent in sheet metal presses, and parts made by "powder metallurgy," a technology for forming metal parts in near-final shape by applying extreme pressure and heat to metal powder.

Finishing.

Many fabrication processes leave "burrs" on the part which must be removed by subsequent operations. In some cases, parts are also washed, painted, polished, or coated.

Assembly.

The finished parts are put together to produce a final product or, alternatively, to produce "subassemblies" which are portions of the final product. In most factories, assembly is moving away from being primarily a manual activity. This phase of manufacturing is receiving increased attention, ranging from design strategies that minimize and simply assembly tasks to automation of the tasks themselves.

Quality Assurance and Control.

There are many quality strategies. They can be divided roughly into those that take place in the design of the product, during fabrication and assembly, and those that take place after a product or subassembly is complete. Quality has been receiving increasing attention in industrial literature and discussion, although the extent to which companies are actually paying more attention to quality on the factory floor is uncertain. There appears to be a movement toward quality assurance (QA) as opposed to quality control (QC) in order to enhance quality and prevent the production of faulty products, as opposed to detecting flaws after production. The Japanese mechatronic approach of doing the job right the first time is slowly catching on in the United States. Strategies for QA range from "quality circles," in which a team of employees helps address production issues that affect quality, to in-process measurement of products as they are manufactured. In the latter, detecting problems in production equipment can sometimes be determined and corrected before the machine makes a bad part. Most complex products are produced with some combination of QA and QC.

Strategies for attaining the more traditional quality control vary widely according to the nature and complexity of the part. The dimensions of mechanical products can be measured, either with manual instruments or with a Coordinate Measuring Machine or laser measurement device; or the product can be compared to one of known quality or to a master gauge. Electronic products can be tested with other electronic devices or probes.

Key Problems in Discrete Part Manufacturing

The following brief outline of the manufacturing process suggests some of the key problems in manufacturing. Underlying each of these problems are the central concerns for any business, those of minimizing cost and risk.

Information Flow.

In any company, small or large, the amount of information that must flow between and among design, manufacturing, and management staff is staggering. In a design process involving several teams of people, how does one make sure that all design and manufacturing personnel are working from the most up-to-date set of plans? How can staff get up-to-date information on the status of a particular batch of parts, or the performance of a particular machine tool or manufacturing department? How can the company keep track of work in process and other inventory?

Coordination.

Beyond merely obtaining information in a timely fashion, the company must use that information to determine how to coordinate its operations effectively. One set of such issues involves coordination of design and production efforts. How can one design products that can be manufactured most efficiently with a given set of tools? How can one minimize the number of parts in order to facilitate assembly? Another set of co-ordination issues arises on the factory floor itself. What is the most efficient way to allocate machines and personnel? How does one adapt the schedule when conditions inevitably change (raw materials do not arrive, production is slower than expected, etc.)?

Efficiency.

Given a large set of choices regarding tools, personnel, and factory organization, a company generally seeks to make the greatest number of products using the fewest resources. How can the company minimize expensive work-in-process inventories? How can manufacturers maximize the percentage of time spent making parts, as opposed to moving them, repairing or setting up machines, and planning? How can the use of expensive capital equipment be maximized? Finally, quality issues within the production process have had a large impact upon efficiency. How can manufacturers maximize the number of products made right the first time, and hence minimize scrap, rework to correct manufacturing errors, and testing?

Flexibility.

Increasingly, issues of flexibility and responsiveness in the manufacturing enterprise are prominent for manufacturers, especially for traditional "mass production" plants. Flexibility is defined here as the range of products and the range of volumes of a specific product which a plant can economically produce. Increased levels of competition, shorter product cycles, and increased demands for customized products are some of the reasons for an emphasis on flexibility. How can the administrative procurement processing time be reduced? How can the turnaround time for design and manufacture of a product be reduced? How can the set-up time for producing a new product be reduced? What is the optimum level of technology for both economy of production and maximum flexibility?

CAD offers improvements in each of these four key areas of manufacturing by applying computerized techniques to control production tools, to gather and manipulate information about the manufacturing process, and to design and plan that process. Further, the use of CAD promises an increase in the degree of control over the enterprise. Many industrialists argue that the more closely manufacturing processes are tied to one another, and the more information is readily available about those processes, the less chance for human error or discretion to introduce unknown elements into the operation. Such control is much harder to realize than it appears in theory. The issue of control is a recurrent theme in this book.

In summary, CIM systems such as CAD can help make factories "leaner" and more responsive, hence, reduce both costs and risks in manufacturing. It is not, however, a panacea for problems in manufacturing. Each factory has different appropriate levels of automation, and there are technical and organizational barriers to implementing CAD most effectively. CAD's capabilities and characteristics from a technical standpoint will be elaborated in the following section.

Computer-Aided Design (CAD)

In the late 1950s and early 1960s, aircraft and automobile companies, whose products are very complex, developed their own software to aid in product design and engineering. Pioneer users such as GM and Boeing were necessarily large firms because early CAD and engineering required the use of expensive mainframe computers. The diffusion of CAD during the 1960s was slow, limited by the cost of hardware and the requirements for extensive engineering and software support. Most early users were defense contractors in the aerospace and electronics industries, where the U.S. Department of Defense (DOD) supported CAD development and its use.

A formal market for the purchase and sale of CAD emerged during the 1970s, due in part to improvements in computer hardware, increased memory capacity, and reduced system cost, and in operating systems which enabled more firms to afford computers for increasingly powerful work. Using microprocessors, mini- and microcomputers made many tasks, including basic two-dimensional CADD, possible without a mainframe computer. The electronics industry, from component manufacturers to computer makers, provided a growing market for CAD systems. Electronics firms were more comfortable than mechanical manufacturing firms with computer-based technology. Their integrated circuit (and printed wiring board) design applications were fundamentally two-dimensional, and therefore well suited to early CAD. Also, the growing complexity of integrated circuits made computer assistance in design increasingly necessary; manual design would require exorbitant amounts of time and manpower. Another early commercial application was in two-dimensional drafting for mechanical design.

During the 1970s, improvements in software and two- and especially three-dimensional CAD fueled a market expansion into mechanical and mapping as well as architecture, engineering, and construction (nuclear power systems) application. Some of these advances stemmed from government-funded efforts, which emphasized aerospace and electronics applications for CAD and the integration of CAD and CAM.

Between 1973 and 1983, the CAD system market grew from under $25 million in annual sales to over $1.7 billion. Hardware and software makers entered the CAD market with specific applications and packaged systems, as shown in Figure 2–5. Firms that entered the CAD market to fill an applications niche typically grew by increasing the variety of CAD applications they could serve. Turnkey vendors, who assembled and installed systems from components made by various sources, also provided training, support, and both standard and custom software. These vendors, led by Computervision, dominated the initial market. They were successful because their customers lacked the technical sophistication to assemble their own systems (but knew when a turnkey system would work for them), and because their reliance on external sources for hardware and other inputs allowed them to incorporate new technology relatively quickly. Accordingly, in addition to system vendors, the CAD industry grew to include groups of hardware and software producers serving both turnkey firms and users directly.

During the mid to late 1970s, the Japanese and European markets (especially those in England, France, Sweden, and Norway) grew rapidly, and markets in less developed countries began to emerge (primarily for mapping applications). U.S. firms dominated the CAD market, both within the United States and abroad, largely because of their perceived software and systems engineering strengths.

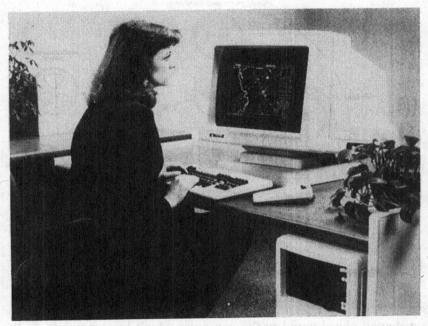

Figure 2-5. Computer-Aided Design System

System Descriptions

In its simpler forms, computer-aided design (CAD) is an electronic drawing board for design engineers and draftsmen. Instead of drawing a detailed design with pencil and paper, these individuals work at a computer terminal, instructing the computer to combine various lines and curves to produce a drawing of a part and its specifications. In its more complex forms, CAD can be used to communicate to manufacturing equipment the specifications and process for making a product (Figure 2-6). Finally, CAD also is the core of computer-aided engineering (CAE), in which engineers can analyze a design and maximize a product's performance using the computerized representation of the product.

The roots of CAD technology are primarily in computer science. CAD evolved from research carried out in the late 1950s and early 1960s on interactive computer graphics, the use of computer screens to display and manipulate lines and shapes instead of numbers and text. SKETCH-PAD, funded by the DOD and demonstrated at Massachusetts Institute of Technology in 1963, was a milestone in CAD development. Users could draw pictures on a screen and manipulate them with a "light pen"—a

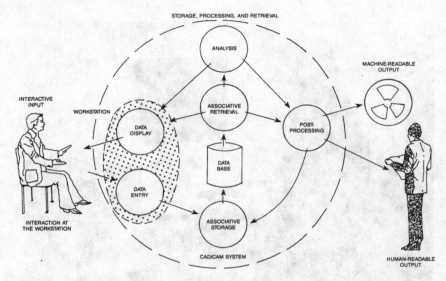

Figure 2-6. Functional Organization of a CAD-CAM System

pen-shaped object wired to the computer which locates points on the screen.

Several key developments in the 1960s and 1970s, including a decrease in the cost of computing power, facilitated the development of CAD technology. Powerful mini- and microcomputers were primarily a result of the integrated circuit chip. Another important technological advance was the development of cheaper, more efficient display screens. In addition, computer scientists began to develop very powerful programming techniques for manipulating computerized images.

How CAD Works

There are various schemes for input of a design to the computer system. Every CAD system is equipped with a keyboard or other input device useful for entering and manipulating shapes. The operator can point to areas of the screen with a light pen or use a graphics tablet, which is an electronically touch-sensitive drawing board. A device called a "mouse" can trace on an adjacent surface to move a pointer around on the screen. If there is already a rough design or model for the product, the operator can "read" the contours of the model into computer memory, and then manipulate a drawing of the model on the screen. Finally, if the part is similar to one that has already been designed using the CAD system, the operator can recall the old design from computer memory and edit the drawing on the screen.

CAD systems typically have a library of stored shapes and commands to facilitate the input of designs. These are the functions performed by a CAD system which can enhance the productivity of a designer or draftsman:

- "Replication" is the ability to take part of the image and use it in several other areas of the design when a product has repetitive features.
- "Scaling," allows CAD to "zoom in" on a small part, or change the size or proportions of one part of the image in relation to the others.
- "Rotation" allows the operator to see the design from different angles or perspectives.

Using such commands, operators can perform sophisticated manipulations of the drawing, some of which are difficult or impossible to achieve with pencil and paper. Repetitive designs, or designs in which one part of the image is a small modification of a previous drawing, can be done much more quickly.

The simplest CAD systems are two-dimensional, like pencil-and-paper drawings. These can be used to model three-dimensional objects if several two-dimensional drawings from various perspectives are combined. For some applications, such as electronic circuit design, two-dimensional drawings are sufficient. More sophisticated CAD systems have been developed in the past few years which allow the operator to construct a three-dimensional image on the screen, a capability which is particularly useful for complex mechanical parts.

Some CAD systems include CAD terminals connected to a central mainframe or minicomputer, although some recently developed systems use stand-alone microcomputers. As the operator produces a drawing, it is stored in computer memory, typically on a magnetic disk. The collection of these digital drawings in computer storage becomes a design data base, readily accessible to other designers, managers, or manufacturing staff.

CAD operators have several output options for their design. Most systems have a plotter, which is capable of producing precise and often multicolor paper copies of the drawing. Some systems can generate copies of the design on microfilm or microfiche for compact storage. Others are capable of generating photographic output. The design is also stored on a computer disk, which is accessible and can be modified as design changes occur.

The CAD systems described above are essentially draftsmen's versions of word processors, allowing operators to create and easily modify an electronic version of a drawing. More sophisticated CAD systems can go beyond computer-aided design and drafting (CADD) in two important ways.

First, such systems allow the physical dimensions of the product, and the steps necessary to produce it, to be processed by computer and electronically communicated to computer-aided manufacturing (CAM) equipment. Some of these systems present a graphic simulation of the machining process on the screen, and guide the operator step by step in planning the machining process. The CAD system can then produce a tape which is fed into the machine tool controller and used to guide the machine tool path. Such connections from CAD equipment to CAM equipment bypass several steps in the conventional manufacturing process:

• They cut down the time necessary for a manufacturing engineer to interpret design drawings and establish machining plans.
• They facilitate process planning by providing a visualization of the machining process.
• They reduce the time necessary for machinists to interpret process plans and guide the machine tool through the process.

Second, these more sophisticated CAD systems serve as the core technology for many forms of CAE. Beyond using computer graphics merely to facilitate drafting and design changes, CAE tools permit interactive design and analysis. Engineers can, for example, use computer graphic techniques for simulation and animation of products, to visualize the operation of a product, or to obtain an estimate of its performance. Other CAE programs can help engineers perform finite element analysis, breaking down complex mechanical objects into a network of hundreds of simpler elements to determine stresses and deformations.

Many analytical functions are dependent on three-dimensional CAD systems which draw the design and perform modeling. The machine can calculate and display such solid characteristics as the volume and density of the object. Solid modeling capabilities are among the most complex features of CAD technology.

Benefits of Computer-Aided Design

The value and benefits of computer-aided design (CAD) are now well recognized by all segments of the manufacturing industry. A few of the significant benefits of utilizing CAD systems are summarized in Tables 2-2 through 2-5.

Table 2-2. Major Computer-Aided Design Benefits

IMPROVED PRODUCTIVITY	Computer-aided design (CAD) systems have been responsible for dramatic productivity increases in many professional engineering activities. The most important of these are:
	— drafting — documentation, — design, — estimating, — order entry, and — manufacturing.
BETTER MANAGEMENT CONTROL	CAD systems have contributed to more cooperative and better-informed management, to and control of:
	— engineering — engineering data distribution, — projects, — production scheduling, — estimating, and — order entry.
INTANGIBLE BENEFITS	Many important benefits of CAD are difficult or impossible to quantify, nevertheless, they contribute in a very real way to the success of technology. The most prominent of these benefits are:
	— standardization of graphics, — standardization of methods, — good quality draftsmanship, — reduced vulnerability to error, — faster response, and — professional development.

Applications

Today's CAD/CAM, CAE products are a mix of host-based, supermini-, mini-, and microbased systems; 16-, 16/32-, and 32-bit processors; local area networks; high and low resolution displays; digitizers, and hundreds of software applications packages, from which is emerging a matrix of applications and price ranges to suit an ever-increasing number of industrial needs.

CAD applications address the specialty areas of mechanical design and manufacturing, electronics design and manufacturing, A-E-C (architecture, engineering, and construction), and mapping (Figure 2-7).

Systems selling for over $40,000 per workstation, which include displays, computers, plotters, and software, are generally host-centered systems that employ a mainframe, supermini-, or minicomputer as host. As

Table 2-3. Intangible Computer-Aided Design Benefits

STANDARDIZATION OF GRAPHICS	Human communication is an important aspect of the workings of an engineering team, and standard graphics reduces the time and effort required for recording and exchanging ideas in clear unambiguous terms. CAD systems enforce standards in a pleasant and positive way by making it easier to comply than to use special symbols or parts.
STANDARDIZATION OF METHODS	By storing preprogrammed procedures for common design and drafting tasks, CAD systems reduce the need to reinvent the wheel and waste creative energies. In addition, standard methods help others to understand what was done when designs are reviewed.
QUALITY	A CAD system can help mediocre and novice draftsmen produce superior quality drawings. Accuracy and penmanship is independent of operator skill, and legibility problems are completely eliminated.
ERROR CONTROL	CAD systems have design rule checking software for many applications, which has proven to be extremely effective for error detection. Because CAD takes the drudgery out of design and drafting, the engineer can concentrate efforts on the work.
FASTER RESPONSE	Results are faster using CAD than by manual methods, increasing the number of engineering options that can be explored in situations where time is critical.
PROFESSIONAL	Computer expertise is an increasingly important skill for professional engineers. Exposure to CAD systems builds knowledge and confidence and provides a strong motivation to learn and use the general computational capabilities of CAD systems.

a rule, they support large-scale applications—for example, solids modeling, finite element modeling, logic simulation, associative data base management—and have superior performance.

Systems in the under-$60,000 category are usually workstation-based, with the system processor built into the workstation, and are frequently limited to drafting applications. The response time is noticeably poorer than in their more expensive counterparts, although performance has

Table 2-4. Management and Control Benefits of Computer-Aided Design

ENGINEERING DATA MANAGEMENT	CAD system data is organized into libraries of associated files. Some have significant capabilities for organizing the information they hold for easy retrieval. These capabilities, together with access control facilities such as passwords and group identifiers, force a measure of management control over engineering data that is otherwise overlooked.
ENGINEERING DATA DISTRIBUTION	CAD systems having telecommunication capabilities provide a valuable means of distributing up-to-the-minute engineering information, giving remote manufacturing plants instant access to the most recently released documentation.
PROJECT MANAGEMENT	Project control charts and critical path diagrams stored in a CAD can be revised daily to give management an up-to-date tool for decision making. In addition, CAD systems can capture other information (such as actual design time used for drafting) useful for controlling the project.
SCHEDULING	Flexible scheduling of machine tools achieves their greatest possible utilization by interfacing the manufacturing data base available from a CAD system with order entry and shop schedule data, management can react and readjust manufacturing schedules when machine tools fail, orders are cancelled, or material is unavailable.
ESTIMATING	When used as an estimating tool, the CAD system ensures all material and labor costs are captured and uniform estimating procedures are followed. In addition, there is much greater control over the engineering data and cost information in use by all estimators at a given time.
ORDER ENTRY	Integrating CAD with order entry can provide greater scheduling flexibility, especially where drawings or manufacturing control tapes must be linked to each other.

improved considerably since vendors have begun to base their workstations on 16- and 32-bit microprocessors.

Systems for under $20,000 are based upon personal computers. Software for personal computer-based systems, initially restricted to two-dimensional drafting, is becoming more sophisticated. Solids modeling, electronics design, and mapping packages have been introduced, and even more sophisticated software will soon become available.

Table 2-5. Productivity Benefits of Computer-Aided Design

DRAFTING	Drawings with recurring features or those that are frequently updated are more efficiently drafted and maintained with a CAD system.
DOCUMENTATION	Bills of material and technical illustrations are quickly produced if they can be derived from data already stored in a CAD system.
DESIGN	Calculations of area, volume, weight, deformation, thermal flux, etc. are best performed by a computer. CAD systems can either perform these calculations themselves or prepare input for larger general purpose computers from graphical data already stored in the CAD system. Also, design tasks that involve fitting together or housing a number of parts are very efficiently done with some CAD systems.
ESTIMATING	The ability of CAD systems to associate, store, and recall graphical and text data has been exploited by engineering estimators. Experience has shown that this approach is more productive than manual methods and captures more cost information.
ORDER ENTRY	Some manufacturers have found that a lot of time can be saved by integrating order entry with their CAD system. Major savings can occur when an order must be tied to specific engineering drawings.
MANUFACTURING	Many CAD systems include software for producing NC tapes and other items used for planning the manufacturing process from information entered and stored in the system during the design phase. This reduces the effort necessary to get a part into production.

Changes in computing hardware are having a great impact upon the CAD industry. The introduction of 32-bit minicomputers (with virtual-memory operating systems) offered improvement over the 16-bit standard, changed the competitive ranking within the industry, and broadened the market. The increase in computing power made minicomputers competitive with mainframes across a variety of CAD applications, such as simulation and solid modeling.

The introduction of low cost, microcomputer-based CAD systems also broadened the CAD market. These CAD systems—generally stand-alone

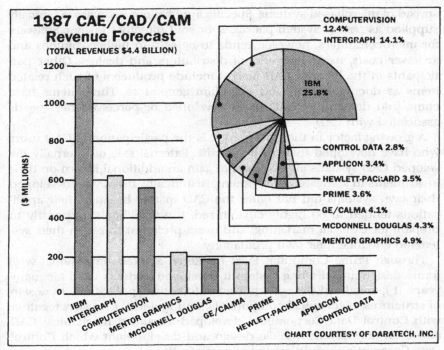

1987 CAE/CAD/CAM Revenue Forecast
(TOTAL REVENUES $4.4 BILLION)

COMPUTERVISION 12.4%
INTERGRAPH 14.1%
IBM 25.8%
CONTROL DATA 2.8%
APPLICON 3.4%
HEWLETT-PACKARD 3.5%
PRIME 3.6%
GE/CALMA 4.1%
MCDONNELL DOUGLAS 4.3%
MENTOR GRAPHICS 4.9%

($ MILLIONS) — 0, 200, 400, 600, 800, 1000, 1200

IBM, INTERGRAPH, COMPUTERVISION, MENTOR GRAPHICS, MCDONNELL DOUGLAS, GE/CALMA, PRIME, HEWLETT-PACKARD, APPLICON, CONTROL DATA

CHART COURTESY OF DARATECH, INC.

Figure 2–7. CAD/CAM Market Segments
Source: Chart courtesy of Daratech, Inc.

workstation units—are less powerful than systems with larger computers, but they make basic CAD available to a larger group of small manufacturers.

Hardware is the largest cost element for CAD systems, but current competition in the CAD market centers on software. While software determines what a system can do, hardware largely determines how fast a task can be done. As CAD system vendors deliver many systems a year, it tends to be uneconomical for them to produce their own hardware. Instead, they rely on a few mass-producing hardware vendors for their equipment. CAD vendors contribute to the product through software development, systems integration, applications engineering, and other support activities. They produce services that accompany the goods they sell. Firms that have entered the CAD market include DEC, UNISYS, Honeywell, Harris, Prime, Data General, Perkin-Elmer, and Hewlett-Packard.

Independent software suppliers have proliferated to meet special applications needs and to meet the growth in demand associated with the spread of microbased systems. Specific applications software is typically supplied as part of system packages, or sold directly to users. Software for microcomputers, however, tends to be sold in higher volumes and at lower costs, using networks of distributors and dealers. Other participants in the broader CAD market include producers of such related items as documentation and microfilm generators. These items have come into demand as CAD users developed or perceived new needs associated with CAD.

A growing factor in the CAD market is the participation of CAD users who have developed their own systems. External sale of internally developed CAD systems allows users to gain an additional return on their investments in software development. Historically, users who developed their own systems did not enter the CAD market because their applications tended to be highly customized. It was difficult and costly to prepare for external marketing, and users preferred to retain their systems to enhance their own profitability.

Through Prime Computer, Ford marketed a three-dimensional wire frame design and drafting system it developed and has used for many years. It generalized the system from automotive applications to design of structures, mechanical components, and systems. Chrysler, together with Control Data Corporation, developed advanced mechanical CAD and CAM software for vehicle design and development which Control Data Corporation would market as part of a line of computer goods and services. Other user-producers include McDonnell-Douglas (Unigraphics), the French firm Dassault (CATIA), and Northrop (NCAD).

U.S. firms continue to dominate both American and foreign markets for CAD systems. Ninety percent of the U.S. CAD market is served by U.S. firms. The international market has undergone considerable merger, acquisition, and, especially, licensing activity. European firms developed important CAD software, but because they lack significant suppliers of CAD hardware their software has been licensed to U.S. firms.

The Japanese role in the CAD market remains focused on hardware. Japanese vendors tend to be computer firms rather than turnkey companies; they sell systems providing U.S. software under license, although they are developing their own software internally and through a government-sponsored consortium.

Anticipated Trends

The CAD market will remain dynamic for the next several years. Industry analysts predict that it will grow between 30 and 50 percent per year; other forecasts for the CAE submarket anticipate even higher rates of growth. Factors such as expected improvements in system capabilities (especially for 3-D modeling), greater ease of use, and reductions in cost for given capabilities will widen the range of customers by size, industry, and application area. Most analysts expect that mechanical applications and CAE systems will become more prominent in the CAD market, reflecting both technological development and the expected spending growth of manufacturers as they recover from the recent recessions. Mapping and facilities management applications are also expected to grow, serving government, utility, and natural resources development customers. CAE has been a major factor in the growth of the custom microchip market. Expected growth in the microchip market overall and the custom share will spur CAE sales.

As the installed base of CAD systems grows, the role of vendor services (for example, software updates, related training) will grow. This growth will reflect in part the growth in sales to smaller firms, which traditionally buy a variety of services they cannot afford to perform themselves. The growing role of services parallels the experience in the computer industry, where service activities and their proportional contribution to revenues increased with the spread of computer systems.

Compared to other types of firms, CAD vendors may be especially well positioned to link CAD and CAM. The design-to-production chain begins with CAD, and CAD firms are already developing systems for modeling production activities and communicating production instructions to other equipment. One company offers systems that program NC machine tools, robots, and coordinate-measuring machines; that design and model manufacturing cells; that design tooling, molds, and dies; and that perform computer-aided process planning. It offers multifunction systems, such as a system for plant design, engineering, construction, and management.

Some vendors are moving away from dedicated CAD terminals in favor of general purpose engineering/professional workstations. These workstations would accommodate not only drafting and design, but also research, software development, and "office automation" functions; they would facilitate shifts in customer activities and software preferences; and lower the risk of hardware obsolescence. Multifunction workstations would aid manufacturing integration, especially when combined with sophisticated data communication systems linking engineering, production, and general corporate data bases. An alternative approach is to market low cost, dedicated CAD workstations which can be linked to mainframe computers for other functions that use a common data base.

CAD Application Thrusts

At the end of 1983 there were an estimated 42,000 CAD workstations in the United States, with aerospace and electronics uses of CAD leading the state of the art. The Boeing Commercial Airplane Company, which began using CAD in the late 1950s, employed the technology extensively in the design of its new generation 757 and 767 aircraft. Boeing uses CAD to design families of similar parts such as wing ribs and floor beams. CAD allows designers to make full use of similarities between parts so that redesign and redrafting are minimized. Moreover, CAD has greatly simplified the task of designing airplane interiors and cargo compartments, which are often different for each plane. Moving seats, galleys, and lavatories are relatively simple with CAD, and the system is then used to generate instructions for the machines which later drill and assemble floor panels according to the layout. Finally, Boeing uses CAD and related interactive computer graphics systems as the basis for computer-aided engineering applications such as checking mechanism clearances and simulating flight performance of various parts and systems.

Computer-aided engineering has also become important in the automobile and aerospace industries, where weight can be a critical factor in the design of products. These industries have developed CAE programs that can optimize a design for minimum material used while maintaining strength.

Applications for the design of integrated circuits are similarly advanced. Very large-scale integrated (VLSI) circuits for example, have become so complicated that it is virtually impossible to track the circuit paths manually and make sure the patterns are correct. There is less need here for geometrically sophisticated CAD systems (integrated circuit designs are essentially a few layers of two-dimensional lines), and more need for computer-aided engineering systems to help the designer cope with the intricate arrangement of circuit pattern. Such CAE programs are used to simulate the performance of a circuit and check it for "faults," as well as to optimize the use of space on the chip.

CAD is also being used in smaller firms; these developments are being spurred by the marketing of relatively low priced turnkey systems— complete packages of software and hardware which, theoretically, are ready to use as soon as they are delivered and installed. Very low cost systems which run on common microcomputers have been introduced, and these have potential uses in a wide variety of firms which otherwise might not consider CAD. The cost of custom developed, specialized systems such as those described above for aerospace and electronics applications is harder to gauge but runs well into the millions of dollars.

The potential advantage of CAD for large as well as small mechanical manufacturing firms is that it facilitates use of previous designs, and allows design changes to be processed more quickly. Because CAD reduces the time necessary for design, it can also improve design by allowing designers to "try out" a dozen or a hundred different variations, where previously they might have been limited to building perhaps three or four prototype models. It also allows many drawings to be rapidly constructed, especially with an experienced CAD operator. Comparisons of design time with CAD range from 0.5 to 100 times as fast as manual systems, with two to six times as fast being typical. Other applications of CAD, though not directly connected to manufacturing, include mapping, architectural drawing and design, graphics for technical publishing, and animation.

Trends and Barriers to Application

There are at least three generations of CAD equipment in use today. The first are the two-dimensional computerized drafting systems mentioned earlier, which streamline the process of drawing and, especially, editing the drawings of parts, plans, or blueprints. The second generation are three-dimensional CAD systems, which allows the user to draw an image of a part using either wireframe models or "surfacing" (displaying the surfaces of objects).

The third generation, more recently available, are the so-called "solid modelers." Such systems (which actually expand 3-D capability) can draw an object in three dimensions and obtain a realistic visualization of the part. Users can rotate, move, and view the part from any angle, and derive performance characteristics. Because the system "constructs" a sophisticated solid model of an object, it can be used to visualize such design issues as component clearance problems. One can even "pull out, a drawer" to make sure it does not hit a cable, for instance.

The increased sophistication of three-dimensional systems greatly improves the ability of such systems to communicate design specifications to manufacturing equipment. There is presently a need for a fourth generation CAD system which offers even more "intelligent" design assistance and can be easily linked to other CAD/CAM systems for manufacturing and management.

Three related themes are evident in current CAD research:

1. improving the algorithms for representing objects using the computer, so that designers can create and manipulate complex objects in an efficient and intuitively clear fashion;
2. adding "intelligence" to CAD systems so that they prevent design errors and facilitate the design process;
3. developing effective interfaces between CAD systems and manufacturing and management.

Improving Algorithms.

Representing and manipulating shapes in computer memory remains a difficult challenge for computer researchers. As the power and complexity of CAD systems increase, their computing needs grow rapidly. One of the problems in manipulating complex shapes with the computer is illustrated by the experimental CAD system used for computer cabinet design at IBM: One of its creators reported that a typical manipulation of a complex object—say, generating an image of the cabinet from a different viewing angle, with all hidden lines removed—might take several minutes of computer processing time. Although the system is still useful, quicker response is clearly needed for the designer to have optimal flexibility from a CAD system. A shorter response time can come from a faster computer or from more efficient ways of representing and manipulating shapes in computer memory.

Much of the current research on CAD involves attempts at more efficient representations. The efficiency of a certain approach also depends on how easy it is to use. A wide variety of schemes are being studied, none of which has a clear overall superiority. One, called "constructive solid geometry," involves assembling images by combining simple shapes, such as blocks, cylinders, and spheres. The other is boundary representation, in which an object is constructed as a set of individual surfaces. One system developed by a group at the University of Utah is based upon "splines": designers manipulate, on the screen, the equivalent of the thin metal strips used in models of boats or planes. They can expand them, curve them, cut them, and so forth to create the model.

There is some concern that not enough time and effort in industry is being devoted to expanding the technologies, particularly the algorithms available for "solid modeling," for example, for true three-dimensional representations of objects. Thus the "experience base" of industries experimenting with three-dimensional systems is very small, and such experience is necessary to refine the systems and determine the needs of manufacturing industries.

Adding "Intelligence to CAD."

In the industry there is much discussion of "smart" CAD systems which would prevent certain operator errors such as the design of an object that could not be manufactured, a case without a handle, or a faulty circuit board. Further, they would facilitate the designer's work by such functions as comparing a design to existing designs for similar objects, and storing data on standard dimensions and design subunits, such as fastener sizes and standard shapes. Smart systems might also increase the ability of CAD systems to simulate the performance of products. There is much concern over "bad design" in industry, and "smart" CAD systems are considered one way to improve the situation.

Though such systems have become rather advanced in electronics applications and offer some hope of becoming more so, there is as yet little in the way of "smart" CAD systems for mechanical applications. A few systems can be programmed to question a designer's choice of certain features that are nonstandard—a 22-mm screw hole in a shop that used only 20-mm and 30-mm holes, for instance. It may soon be possible to use an "expert" system for developing a "smart" CAD system.

CAD as Part of Computer-Integrated Systems

The most important research would involve connecting computer-aided design to other computerized systems in the factory. Such connections would mean that design information could be forwarded directly to machine tools that make the parts, that designers could draw on previous designs as well as on data on their performance and cost, and that designers would have up to date information on the manufacturability and cost of their designs.

There has been significant progress toward interfaces between CAD devices. The Initial Graphics Exchange Specification, developed at the National Institute of Standards and Technology, allows different CAD systems to exchange data. Interfaces between computer-aided design systems are becoming easier, but more progress is needed to allow CAD and CAM systems to communicate. These devices can be wired together into a computer network, but establishing an effective data-driven interface requires sophisticated software to manipulate manufacturing information to make it useful for designers.

Movement toward design-manufacturing connections is impeded by a strong tradition of separatism among design engineers and manufacturing engineers. A common description of the relationship is, "The design engineer throws the set of drawings over the wall to manufacturing." There is evidence that such barriers are beginning to break down, slowly, as the need for communication has become apparent, and as engineering schools have begun to broaden the connections between design and manufacturing curricula.

There are many research efforts whose ultimate goals include such connections between CAD and other manufacturing systems. These research programs include the Air Force's Integrated Computer-Aided

Manufacturing project, as well as the National Institute of Standards and Technology Advanced Manufacturing Research Facility, and a joint West German/Norwegian effort. The heart of the latter effort is an attempt to use a very advanced geometric modeling system developed by the Technical University of Berlin as the basis for developing software that would allow design to be connected to all aspects of the manufacturing process. In addition, users of CAD/CAM, such as Westinghouse, GE, and IBM, are also working on interface issues.

Pace of Technology Transfer

The rate of growth in use of CAD in the United States, known as the "diffusion" of the technologies, depends on factors both in the larger economy and at the level of individual firms and products. Some of the more general factors include availability of capital and skilled labor, international competition, and the amount of attention U.S. firms devote to improvements in manufacturing processes. The last factor may be the most critical. Manufacturing engineering in the United States has been largely neglected both in engineering schools and in industry.

Prompted in part by international competition, however, the mood among American industrialists seems to be changing. Increasingly, established management practices are being questioned in conferences and industry journals, and many industrial managers are closely examining improvements in manufacturing processes, particularly CAD, CAM, CIM, and robots. The extent to which this change in mood will effect lasting and significant change in manufacturing, however, is uncertain. Many management specialists believe that such lasting change must include discarding powerfully entrenched habits in industry, particularly financially-oriented management strategies that discourage risk-taking and downplay quality relative to cost.

In addition to these more general questions, a large number of factors come into play when an individual firm chooses to use or not to use CAD. Some of the technical factors include: the applicability of the technology to the problem at hand, which tends to vary according to the particular manufacturing processes used in each factory; the range of tasks to which a given tool can be applied; the cost of customization, particularly for new technologies where few standards exist and almost every application is a prototype; the ease of use of the tool; the reliability of the equipment; the compatibility of CAD with machines already in place; and finally, the capacity of different CAD systems for upgrading and expansion.

Organizational factors can also have a significant effect on a firm's automation decision. Previous experience with automation is a key factor in successful applications, and industry observers report that many unsuccessful attempts to use CAD have been due to premature jumps into complex systems. There can also be substantial resistance to change on the part of workers or management. Many manufacturers report, however, that production workers tend to accept technological changes such as automation, while strong resistance tends to come from middle managers who fear CAD will diminish their degree of control or eliminate their jobs.

The Future

In recent years, CAD has responded to the industrywide demand for interactive graphics systems that can do more things faster and for a smaller capital investment. The memory capacity and circuit densities of the electronic components of CAD systems have quadrupled every four years, enabling users to store and execute increasingly sophisticated programs in less time at a lower cost per operator. The speed and capacity of computer peripherals have also been accelerating. These trends will intensify through the 1980s. The systems of the future will feature not just more CAD capabilities, but more nongraphical capabilities and computing power as well, supported by far greater computer memory.

Providing faster response for today's CAD users requires tight coupling (high speed communication lines) between the central processing equipment and the workstations. This seriously restricts the use of CAD, since many companies with widely distributed operations are anxious to provide CAD facilities to their remote locations. Small design offices scattered around the country would like to service local clients. Each of these design offices would like to access a common CAD data base.

One can expect a new family of CAD systems to emerge in which workstations are considerably more intelligent than currently, and provide a great deal of local picture manipulation. The central system need not be accessed frequently for the moment-to-moment design and drafting functions. However, all systems will need to be networked and have access to a common data base so that the benefits of standardization can be realized.

Of course, this can be done today in principle, but the cost of a fully intelligent workstation is high. With the cost of computation coming down and other price factors becoming more favorable toward peripherals, display devices, and so on, one can expect lower costs for intelligent workstations in the future. This, of course, implies that the same workstations might be used on a stand-alone basis for those companies which can not afford a larger, more sophisticated system.

Within a few years, all the intelligence and "computer power" now resident in today's most advanced CAD systems may be available at every terminal in the system, yet at a per-terminal cost modest enough to allow companies to give every designer and draftsman a terminal. Thus, engineers will be able not only to design a part at the terminal, but to conduct various engineering and stress analysis tests on that product.

As the cost of computation comes down, the capability of the central CPU, enhanced by distributed processing, can be expected to increase considerably. The next decade will see the emergence of much lower-cost mass storage, so that it will be possible to store tens of thousands of drawings "on-line," instead of the thousand or so drawings characteristic of present systems. The ultimate goal will be a totally automated design and production process. More and more intelligence and diagnostics will be built into the systems, including remote diagnostics connected through telecommunications to a central service center. Systems will have redundancy so that if a portion of the system does not work or breaks down, its function can be shifted to another network.

In display technology, small plasma panels are now available, but are still expensive. There will be continuing development of small desktop flat-panel displays—a touchstone of graphics since the mid-1950s. Various kinds of flat-panel matrix units are advancing, with plasma and liquid crystal displays looking the most promising.

Interest continues in large-screen displays, particularly in the expanding market for business information systems. Standard CRT projection technology should remain competitive with new techniques based upon various kinds of light valves, such as liquid crystal. Lasers may again emerge as a technology for large-screen projection.

Because of continuing reduction in logic and memory costs, raster displays will continue to be the dominant display technology through the 1980s. Low resolution (500 lines or less) systems will continue to decline in cost. By the end of the decade a low resolution color system should cost about what a standard television set costs today. There will also be continuing advances in higher resolution systems. Thousand-line systems should become fairly common, with an intermediate 2,000-line system, and near the end of the decade the technology should begin to push into 4,000-line systems.

Terminals.

From a device standpoint, one can expect a series of system enhancements. Color and color contrasts have become significant improvements in CAD. New techniques which make brighter images that change faster are being developed. New techniques for communicating with the image by simply touching the screen surface, rather than using the electronic pen, are now available and can be expected in future systems.

Software.

A major software trend, as graphics move into more applications in the 1980s, will be to make CAD software more user friendly to satisfy the nontechnical, nonprogrammer user. CAD software will change in form to the point where application and system software will be furnished, as plug-in firmware semiconductor modules, by hardware manufacturers and independent software suppliers.

The use of true geometric modeling will become more common in computer-aided design by the end of the decade. In CAD and other graphics systems, there is a continuing movement toward turnkey systems. As well-defined applications develop, specially designed terminals will emerge containing all the applications software, either as conventional software or as firmware, and there will be custom designed operator input elements specifically oriented to those solutions.

Interactive Graphics and Voice Recognition

The video disk is becoming an increasingly important part of CAD. A cost-effective read/write capability is developing, and interactive systems built around read/write video disks will be used.

As applications develop in which a user wants graphics images, three-dimensional capabilities become very important. But where accurate assessments need to be made, the user can probably draw better conclusions from orthographic, two-dimensional representations. In spite of the spectacular imagery that is possible, holography will not become a factor in computer graphics over the next decade.

Finally, the techniques of voice recognition will become more refined over the next decade. One can expect to see terminals that recognize speech input become part of the CAD environment.

Computer-Aided Manufacturing (CAM) Technologies

Computer-aided manufacturing (CAM), a widely used term in industrial literature, has various meanings; here it refers simply to those types of CAM systems used primarily on the factory floor to help manufacture products. The following sections on numerically controlled machine tools, flexible manufacturing systems, and automated materials handling systems provide functional descriptions of CAM tools.

Numerically Controlled Machine Tools

Numerically controlled (NC) machine tools are devices that cut a piece of metal according to programmed instructions concerning the desired dimensions of a part and the steps for the machining process. These devices consist of a machine tool, specially equipped with motors to guide the cutting process, and a controller that receives numerical control commands (see Figure 2–8).

The U.S. Air Force developed NC technology in the 1940s and 1950s, in large part to help produce complex parts for aircraft that were difficult to make reliably and economically with a manually guided machine tool.

Figure 2–8. A Numerically Controlled Machine Tool

Source: Bridgeport Machines, Inc.

How They Work.

Machine tools for cutting and forming metal are the heart of the metalworking industry. Using a conventional manual machine tool, a machinist guides the shaping of a metal part by hand, moving either the workpiece or the head of the cutting tool to produce the desired shape of the part. The speed of the cut, the flow of coolant, and all other relevant aspects of the machining process are controlled by the machinist.

In ordinary NC machines, programs are written at a terminal which in turn punches holes in a paper or mylar plastic tape. The tape is then fed into the NC controller. Each set of holes represents a command, which is transmitted to the motors guiding the machine tool by relays and other electromechanical switches. Although these machines are not computerized, they are programmable in the sense that the machine can easily be set to making a different part by feeding it a different punched tape; and they are automated in that the machine moves its cutting head, adjusts its coolant, and so forth, without direct human intervention. However, most of these machines still require a human operator, though in some cases one person operates two or more NC machine tools. The operator supervises several critical aspects of the machine's operation:

1. The operator has override control to modify the programmed speed (rate of motion of the cutting tool) and feed (rate of cut). These rates will vary depending on the type of metal used and the condition of the cutting tool;
2. The operator watches the quality and dimensions of the cut and listens to the tool, ideally replacing worn tools before they fail; and
3. The operator monitors the process to avoid accidents or damage such as a tool cutting into a misplaced clamp or a blocked coolant line.

Typically, NC programs are written in a language called APT (automatically programmed tools), which was developed during the initial Air Force research on NC. A number of modified versions of APT have been released in the last decade, and some of these are easier to use than the original; the essential concept and structure of the numerical codes, however, has remained the same. In large part because of the momentum it gained from its initial DOD support, APT has become a de facto standard for NC machine tools.

Since 1975, machine tool manufacturers have begun to use microprocessors in the controller, and some NC machines—called computerized numerically controlled (CNC) machines—come equipped with a dedicated minicomputer. CNC machines are equipped with a screen and keyboard for writing or editing programs at the machine; closely related to CNC is *direct numerical control* (DNC), in which a larger mini- or mainframe computer is used to program and run two or more NC tools simultaneously. As the price of small computers has declined over the past decade, DNC has evolved both in meaning and concept into *distributed numerical control,* in which each machine tool has a microcomputer of its own, and where the systems are linked to a central controlling computer. One of the advantages of such distributed control is that the machines can often continue working for some time even if the central computer "goes down."

The machining processes are essentially the same in all types of NC machine tools; the difference is in the sophistication and location of the controller. CNC controllers allow the operator to edit the program at the machine, rather than sending a tape back to a programmer in a computer room for changes. The tape punchers and readers and the tape itself have been notable trouble spots; by avoiding the use of paper or mylar tape, CNC and DNC machines are substantially more reliable than ordinary NC machines. In addition, CNC and DNC machines, through their computer screens, offer the operator more complete information about the status of the machining process. Apart from those features associated with CNC and DNC, some NC tools are equipped with a feature called "adaptive control," which tries to optimize the rates of cut automatically to produce the part as fast as possible, while avoiding tool failure.

Applications.

The diffusion of NC technology into the metalworking industry proceeded very slowly in the 1950s and 1960s, though it has accelerated somewhat over the past ten years. In 1983, numerically controlled machine tools represented only 4.7 percent of the total population, although this figure may be somewhat misleading: the newer NC machine tools tend to be used more than the older equipment, and firms often keep old equipment even when they buy new machines. Some industry experts have estimated that as many as half of the parts made in machine shops are made using NC equipment. Nevertheless, the applications still tend to be concentrated in large firms and in smaller subcontractors in the aerospace and defense industries.

The U.S. machine tool population is significantly older than that of most other countries, and this situation, suggesting relatively low levels of capital investment, has been a source of concern for many in industry and government. In 1983, for the first time in several decades the percentage of metal-cutting tools less than ten years old increased by 3 percent, although the percentage of metal-forming tools less than ten years old remained at an all time low of 27 percent.

The U.S. government has encouraged diffusion of NC technology, which has moved beyond the aerospace industry, but not as fast as most observers expected. There are several reasons for the relatively slow diffusion of NC technology. These include high capital cost for an NC machine (perhaps $80,000 to $150,000 and up, as opposed to $10,000 to $30,000 for a conventional machine tool). In addition, the successful application of NC machine tools requires technical expertise that is in short supply in many machine shops. Training is also a problem, as some users report requiring as long as two years to get an NC programmer up to speed, while small machine shops typically do not have the resources or expertise to train staff to use or maintain computerized equipment. Finally, according to one source, APT proved to be too complicated for most users outside the aerospace industry. Most machine jobs could be specified in a considerably less complex world.

In spite of the roadblocks to implementing NC, there are some clear advantages. Intricate shapes such as those now found in the aerospace industry are nearly impossible for even the most experienced machinist using conventional machine tools. With NC, the parts can be more consistent because the same NC program is used to make the part each time it is produced. A manually guided machine tool is more likely to produce parts with slight variations, because the machinist is likely to use a slightly different procedure each time he or she makes a part. This may not be a concern in one of a kind or custom production, but it can create problems in batch production. The advantages in consistency brought by NC are seen by many manufacturers as an increase in their control over the machining process.

NC machines tend to have a higher "throughput" than conventional machine tools and hence are more productive. They operate (that is, cut metal) more of the time than do conventional machine tools because all the steps are established before the machining begins and are followed methodically by the machine's controller. Further, on a complex part that takes more than one shift of machining on a conventional machine tool, it is very difficult for a new machinist to take over where the first left off. The part may remain clamped to the machine and the part and machine tool may lie idle until the original machinist returns. On NC machines, operators can substitute for each other relatively easily, allowing the machining to continue uninterrupted.

As discussed previously, the capability of guiding machine tools with

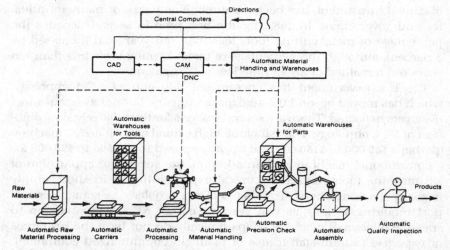

Figure 2-9.   Block Diagram of a Flexible Manufacturing System. Reprinted with the permission of Reston Publishing Company, Prentice-Hall from *Robotics and Automated Manufacturing* by Richard Dorf.

numeric codes opens up possibilities for streamlining the steps between design and production. The geometric data developed in drawing the product on a CAD system can be used to generate the NC program for manufacturing the product.

Flexible Manufacturing Systems

Manufacturing integration, as proposed in the mechatronic factory of the future, will interconnect all the tools first into autonomous workstations, then the workstations into flexible cells, and then the cells into a composite computer-integrated factory. The key is integration at each level all under central process control. The level where raw materials become products, in contrast to parts or assemblies, is in the flexible manufacturing cell. An FMS, shown in Figure 2-9, is a unit capable of producing a range of discrete products with a minimum of manual intervention. It consists of production equipment cells or workstations (machine tools or other equipment for fabrication, assembly, or treatment) linked by a materials handling system to move parts from one workstation to another, and it operates as an integrated system under full programmable control.

An FMS is often designed to produce a family of related parts, usually in relatively small batches, in many cases less than 100 and even as low as one. Most systems appropriately considered to be FMSs include at least four workstations, while some have up to 32. Smaller systems of two or three machine tools served by a robot, which are also called flexible manufacturing systems in some circumstances, are more appropriately termed flexible "machining cells."

How an FMS Works.

Using NC programs and (often) computer-aided process planning, workers develop the process plan (that is, the sequence of production steps) for each part that the CIM system produces. Then, based on inventory, orders, and computer simulations of how the FMS could run most effectively, the FMS managers establish a schedule for the parts that the FMS will produce on a given day. Next, operators feed the material for each part into the system, typically by clamping a block of metal into a special carrier that serves both as a fixture to hold the part in place while it is being machined, and as a pallet for transporting the workpiece. Once loaded, the system itself essentially takes over. Robots, conveyors, or other automated materials handling devices transport the workpiece from workstation to workstation, according to the process plan. If a tool is not working, many systems can reroute the part to other tools that can be substituted for the defective unit.

Machine tools are not the only workstations in an FMS; other possible stations include washing or heat-treating machines and automatic inspection devices. While most current FMSs consist of groups of machine tools, other systems anticipated or in operation involve machines for grinding, sheet metal working, plastics handling, and assembly.

The amount of flexibility necessary to deserve the label "flexible" is arguable. Some FMSs can produce only three or four parts of very similar size and shape, for example, three or four engine blocks for different configurations of engines. One FMS expert argues, however, that in the current state of the technology, a system that cannot produce at least 20 to 25 different parts is not flexible. Indeed, some are being designed to manufacture up to 500 parts.

The essential features that constitute a workable "part family" for an FMS are:

A common shape.   In particular, prismatic (primarily flat surfaces) and rotational parts cannot be produced by the same set of machines.

Size.   An FMS will be designed to produce parts of a certain maximum size, for example, a 36-inch cube. Parts that are larger or very much smaller may not be handled.

Material. Titanium and common steel parts cannot be effectively mixed, nor can metal and plastic.

Tolerance. The level of precision necessary for the set of parts must be in a common range.

Applications.

For a manufacturer with an appropriate part family and volume to use an FMS, the technology offers substantial advantages over stand-alone machine tools. In an ideal FMS arrangement, the company's expensive machine tools work at near-full capacity. Turnaround time for manufacture of a part is reduced dramatically because parts move from one workstation to another quickly and systematically, and computer simulations of the FMS help determine optimal routing paths. Most systems have some redundancy in processing capabilities, and this can automatically reroute parts around a machine tool that is down. Because of these time savings, work-in-process inventory can be drastically reduced. The company can also decrease its inventory of finished parts, since it can rely on the FMS to produce needed parts on demand.

Finally, FMS can reduce the "economic order quantity"—the batch size necessary to justify set-up costs—for a given part. When a part has been produced once on an FMS, set-up costs for later batches are minimal because process plans are already established and stored in memory, and materials handling is automatic. In the ultimate vision of an FMS, the machine could produce a one-part batch almost as cheaply as it could produce 1,000 in cost per unit. While there are, in practice, unavoidable set-up costs for a part the FMS's capability, to lower the economic order quantity is particularly useful in an economy in which manufacturers perceive an increased demand for product customization and smaller batch sizes.

A midwestern agricultural equipment manufacturer, for example, uses an FMS to machine transmission case and clutch housings for a family of tractors. The company had considered "hard automation"—a transfer line—to manufacture the parts but expected a new generation of transmissions within five years, which would render the transfer line obsolete. They chose an FMS instead because it could be more easily adapted to other products. In the system, a supervisory computer controls 12 computerized machining centers and a system of chain-driven carts that shuttle the fixtured parts to the appropriate machines. The supervisory computer automatically routes parts to those machines with the shortest queue of workpieces waiting and can reroute parts to avoid a disabled machine tool. About a dozen employees operate and maintain the system during the day shift, and there are even fewer people on the other two shifts. The system is designed to produce nine part types in almost any sequence desired. (It is, therefore, rather inflexible according to the current state of the art.) This system, in fact, was one of the earliest FMSs of substantial size to be designed. It was ordered in 1978, but not fully implemented until 1981.

Despite the advantages claimed for FMS, there are relatively few systems installed. Observers estimate that there are 50 such systems in Japan, 20 each in Western and Eastern Europe, and 50 in the United States. The reasons for this scarcity of application include the complexity, newness, and cost of the systems. One U.S. manufacturer estimated that an FMS costs $600,000 to $800,000 per machining workstation, with a minimum expenditure of $3 million to $4 million. In addition, the in-house costs of planning for installation of an FMS—a process which often takes several years—are likely to substantially increase the investment.

Automated Materials Handling Systems

Automated materials handling (AMH) systems store and move products and materials under computer control. Some AMH systems are used primarily to shuttle items to the work areas or between workstations on automated carts or conveyors, as shown in Figure 2–10. Automated storage and retrieval systems (AS/RS) are another form of automated materials handling, essentially comprising an automated warehouse where parts are stored in racks and retrieved on computerized carts and lift trucks.

How AMH Systems Work.

There are a wide variety of formats for automated materials handling. They include conveyors, monorails, tow lines, motorized carts riding on tracks, and automated carriers that follow wires embedded in the floor of the factory. Each AMH system is unique, and each is designed for the materials handling needs of a particular factory. The common characteristics of these devices is that they are controlled by a central computer.

There are three general applications for AMH. The first is to shuttle workpieces between stations on a CIM system. In this case, the AMH system operates on commands from the CIM controller. For example,

Figure 2-10. Automated Material Handling System at Hughes Aircraft Company

Source: Hughes Aircraft Company

when the controller receives a message that a machine tool has finished work on a certain workpiece, the controller orders the AMH system to pick up the workpiece and deliver it to the next workstation in its routing. The materials handling portion of CIM is one of its trickiest elements; parts transport needs tend to be logistically complicated, and the AMH system must place the parts accurately and reliably for machining.

Many AMH systems, such as conveyors or tow chains are serial in nature, that is, there is only one path from Point A to Point B. This has caused CIM systems to cease operating when a cart becomes stuck or a critical path becomes unusable. CIM designers have responded to this problem by designing AMH systems with backup paths, and by using systems such as the wire-guided vehicle mentioned earlier, which can be routed around disabled carts or other obstacles.

The second major application of AMH is for transporting work in process from one manufacturing stage to the next within a factory. This application is similar in concept to AMH use for a CIM system, although serving an entire factory is more complex. There is more area to cover, more potential obstacles and logistical difficulties in establishing paths for the AMH carriers, and a wider range of materials to handle. For this reason, whole-factory AMH systems are not yet widely used. However, a few years ago, General Motors agreed to purchase automatic guided vehicles from Volvo which allow automobiles to proceed independently through the plant while being assembled. The "robot carts" can be programmed to stop at appropriate workstations, and the cart system essentially replaces an assembly line. Volve uses about 2,000 of the carts in its own plants in Europe, and Fiat also uses such carts in Italy.

The final application for AMH is in automated storage and retrieval systems. These storage rack systems are often very tall in order to conserve space and to limit the number of automatic carrier devices needed to service the facility. In many cases the structure housing the AS/RS is a separate building adjacent to the main factory. Design of an AS/RS depends on the size of the products stored, the volume of material to be stored, and the speed and frequency of items moving in and out of the system. Advocates of AS/RS cite advantages for the system, as compared to nonautomated systems, which include reduced land needs for the plant, fewer (but more highly trained) staff, more accurate inventory records, and reduced energy use.

Applications.

In theory, AMH systems can move material quickly, efficiently, and reliably, while also keeping better track of the location and quantities of the parts by use of the computer's memory (thus avoiding much paperwork). They can therefore minimize loss of parts in a factory, which is a common problem in materials handling.

Deere & Co., for example, uses an extensive AS/RS to store materials and inventory at one of its tractor plants. The system's computerized controller keeps track of the products stored on the shelves, and workers can order the system to retrieve parts from the shelves by typing commands at a computer terminal. After they are retrieved from the AS/RS, the parts can be carried automatically by overhead conveyors to the desired location within the plant complex.

IBM's Poughkeepsie plant has developed an AMH conveyor cart system for transporting a 65-pound computer subassembly fixture between assembly and testing stations. The manufacturing manager reports that the decision to adopt this system was prompted by logistical difficulties in keeping track of many such fixtures among a great variety of workstations, as well as by worker health problems related to transporting the fixtures manually. AMH systems often have reliability problems in practice. According to a Deere & Co. executive, for example, Deere's AS/RS was systematically reporting more engines stored on the racks than other records indicated. After long weeks of searching for the problem, the plant staff finally found the culprit: a leak in the roof was allowing water to drip past the photocell that counted the engines as they were stored. Each drip, in essence, became an engine in the computer's inventory. Although Deere's experience is not widely applicable to AS/RSs, the notion that AMH systems present unexpected logistical and mechanical problems does seem to be generally accurate.

Even though these systems are key aspects of computer-integrated manufacturing systems, materials handling has long been a neglected topic in industrial research. Materials handling system manufacturers have only recently "caught up" to other industrial systems in level of sophistication, and few companies have so far installed sophisticated AMH systems. Because of this relative lack of sophistication, materials handling for CIM, especially for a complex application such as delivery of multiple parts to an assembly station, may be one of the biggest problems facing integrated automation.

Other CAM Equipment

There are several other kinds of CAD/CAM equipment used in manufacturing. Described in brief, they include:

Computer-aided inspection and test equipment.

For mechanical parts, the most prominent such device is the coordinate measuring machine, which is a programmable device capable of automatic and precise measurements of parts. A great variety of inspection and test equipment is also used for electronic parts. IBM's Poughkeepsie plant, mentioned above, performs the vast majority of its testing of microprocessor modules with automatic devices built in-house. In addition, robots can be used as computer-aided inspection and test devices—several two-armed, gantry-style robots are used at IBM to test the wiring for computer circuit boards. In the test, thousands of pairs of pins on the circuit board must be tested to make sure that they are correctly wired together. Each arm of the robot is equipped with an electronic needlelike probe, and by touching its probes to each pair of pins and passing an electronic signal through the probes, the robot's control computer can determine whether the circuit board's wiring is okay.

Electronics assembly.

Increasingly, programmable equipment is used to insert components —resistors, capacitors, diodes, and so on—into printed circuit boards. One such system, called Mini-Semler™ and manufactured by Control Automation, is capable of inserting 15,000 parts per hour.

Process control.

Programmable controllers (PCs) are being used extensively in both continuous-process and discrete-manufacturing industries. PCs are small, dedicated computers which are used to control a variety of production processes. They are useful when a set of electronic or mechanical devices must be controlled in a particular logical sequence, as in a transfer line where the conveyor belt must be sequenced with other tools, or in heat treatment of metals in which the sequence of steps and temperature must be controlled very precisely. Until the late 1960s, PCs were composed of mechanical relays, and were "hard-wired"—one had to physically rewire the device to change its function or the order of processes. Modern PCs are computerized and can typically be reprogrammed by plugging a portable computer terminal into the PC. A computerized PC is not only more easily reprogrammed than a hard-wired device, but is also capable of a wider range of functions. Modern PCs, for example, are often used not only to control but also to collect information about production processes. PCs and numerical control devices for machine tools are very similar in concept, essentially, NCs are a specialized form of PC designed for controlling a machine tool.

Tools and Strategies for Manufacturing Management

Several kinds of computerized tools are becoming available to assist in management and control of a manufacturing operation. The essential common characteristic of computerized tools for management is their ability to manipulate and coordinate "data bases"—stores of accumulated information about each component of the manufacturing process. The ability to gain access to these data bases quickly and effectively is an extraordinarily powerful capacity for management. What was a chaotic and murky manufacturing process can become much more organized, and its strengths and weaknesses grow more apparent. The following pages describe some of these tools.

Data-Driven Management Information Systems (DDMIS)

Manufacturers use and store information on designs, inventory, outstanding orders, capabilities of different machines, personnel, and costs of raw materials, among other things. In even a modestly complex business operation, these data bases become so large and intricate that complex computer programs must be used to sort the data and summarize it efficiently. Management information systems (MIS) perform this function, providing reports on such topics as current status of production, inventory and demand levels, and personnel and financial information. Before the advent of powerful computers and management information systems, some of the information now handled by MIS was simply not collected. In other cases, the collection and digestion of the information required dozens of clerks. Beyond saving labor, MIS bring more flexible and more widespread access to corporate information. For example, with just a few seconds of computer time, a firm's sales records can be listed by region for the sales staff, by dollar amount for the sales managers, and by product type for production staff. Perhaps most importantly, the goal for MIS is that the system be so easy to use that it can be used directly by top-level managers.

Computer-Aided Planning

Computer-aided planning systems sort the data bases for inventory, orders, and staff, and help factory management schedule the flow of work in the most efficient manner. Manufacturing resources planning, perhaps the best-known example of computer-aided planning tools, can be used not only to tie together and summarize the various data bases in the factory, but also to juggle orders, inventory, and work schedules, and to optimize decisions in running the factory (Figure 2–11). In some cases these systems include simulations of the factory floor so as to predict the effect of different scheduling decisions. Manufacturing resource planning systems have applicability for many types of industry in addition to metalworking.

Another kind of computer-aided planning tool is computer-aided process planning (CAPP), used by production planners to establish the optimal sequence of production operations. There are two primary types of CAPP systems, variant and generative.

The variant type, which represents the vast majority of such systems currently in use, relies heavily on group technology (GT). In GT, a manufacturer classifies parts produced according to various characteristics, such as shape, size, material, presence of teeth or holes, and tolerances.

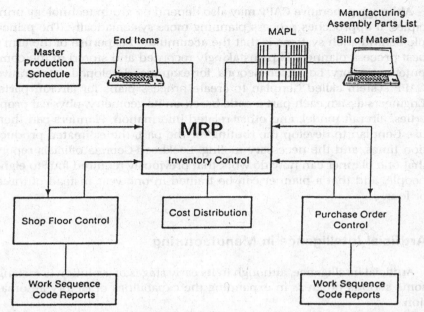

Figure 2-11. Block Diagram of a Manufacturing Resources Planning System. Reprinted with the permission of Reston Publishing Company, Prentice-Hall from *Robotics and Automated Manufacturing* by Richard Dorf.

In the most elaborate GT systems, each part may have a 30- to 40-digit code. GT makes it easier to systematically exploit similarities in the nature of parts produced and in machining processes to produce them. The theory is that similar parts are manufactured in similar ways. So, for example, a process planner might define a part, using GT classification techniques, as circular with interior holes, 6-inch diameter, 0.01-inch tolerance, and so forth. Then, using a GT-based CAPP system, the planner could recall from computer memory the process plan for a part with a similar GT classification and edit that plan for the new, but similar, part.

Generative process planning systems, on the other hand, attempt to generate an ideal routing for a part based on information about the part and sophisticated rules about how such parts should be handled, together with the capabilities of machines in the plant. Unlike process plans in variant systems, therefore, generative systems produce optimal plans.

A variant system uses as its foundations the best guesses of an engineer about how to produce certain parts, so that the variants on that process plan may simply be variations on one engineer's bad judgment.

Although generative CAPP may also depend on group technology principles, it approaches process planning more systematically. The principle behind such systems is that the accumulated expertise of the firm's best process planners is painstakingly recorded and stored in the computer's memory. Lockheed-Georgia, for example, developed a generative CAPP system called Genplan to create process plans for aircraft parts. Engineers assign each part a code based on its geometry, physical properties, aircraft model, and other related information. Planners can then use Genplan to develop the routing for the part, the estimated production times, and the necessary tooling. Lockheed-Georgia officials report that one planner can now do work that previously required four to eight people, and that a planner can be trained in one year instead of three or four.

Artificial Intelligence in Manufacturing

Artificial intelligence, although in its early stages of evolution is making some serious progress in expanding the capabilities of factory automation.

The direction of manufacturing is toward more automation in the entire process, from the concept planning using computer-aided design (CAD), to the factory process planning and production phases using computer-aided manufacturing (CAM). The success of such a system implies production components that are capable of being complemented with artificial intelligence components to make them perform better.

Expert systems, at the present time, are best employed where there is a narrow application that can be well defined by a source expert who can identify rules and the criteria for applying those rules. And since the production line is by definition a well-organized and well-planned place, what better place to capitalize on this new tool.

Robots, currently under computer control are directed by software through a preplanned process allowing for no deviations from that plan. Although efficient, the robot is inflexible when something happens in its area of influence not known or predicted in advance. Newer robots are relying more on sensory input to allow the tool more autonomy in its operation. These additional forms of information retrieval are beginning to excite the artificial intelligence community into developing scenarios where a robot might perform even more freely and have even more autonomy in its work area. The implications are extremely important to the automated manufacturing process. The performance of a humanlike intellect housed in an industrial framework impervious to the dangers of the workplace is truly the best combination for this worker.

While artificial intelligence will play an ever-increasing role in the production robot, the more near-term applications of those expert systems will be in the product design specification, planning, and related chores. These parts of the production process are already being addressed by CAM but with the application of artificial intelligence to the use of CAM, the user can benefit from the experience of a resident expert that will not permit certain design mistakes to occur. It could be a rather simple process to install an expert system that measured mechanical tolerances as a system was being created, making sure that the designer could not make the mistake of building a system that was more reliable than the components making it up. Another example might be the expert system that could perform a worst case design given the performance bounds of the system and the overall cost limitation imposed to make the product feasible.

Imagine a tool that could provide up-front guidance to a CAD designer on the feasibility of his design given a variety of variables including size, weight, cost, reliability, environmental constraints, and then after designing such a system, provide a parts list and process plan for making that system in the factory. The idea seems far out but the tools are already available for other jobs; all that is needed is the motivation and the emphasis to make it happen here, too. There are several expert systems already in use to aid the manufacturing process. Three of these are listed below.

1. IMACS. A field prototype expert created by Digital Equipment Corporation (DEC) assists managers with capacity planning and inventory management. IMACS takes customer orders, generates build plans, and later uses that plan to monitor the implementation of the plan.
2. ISIS. Developed by Carnegie-Mellon it constructs factory job shop schedules, selects the sequence of operations for the job, determines start and end times, and assigns resources to each operation. It also acts as an intelligent assistant by helping plant schedulers maintain schedules and identify decisions that are liable to result in less than optimum outcomes.
3. PTRANS. Developed jointly by DEC and Carnegie-Mellon it helps control the manufacture and distribution of DEC's computer systems. It uses customers orders and information about plant activities to develop a plan to assemble and test the ordered systems. PTRANS monitors the progress of the technicians implementing the plan, diagnoses problems and suggests solutions, and predicts shortages or surpluses.

As sensors become more readily available and cost effective for integration into more factory tools, especially robots, it is certain that intelligent software will be devised to allow them to make intelligent decisions based on those sensory inputs. The missing link at this point is the technology to perform pattern recognition on the sensor data in a reliable and consistent manner. This field of endeavor is being pursued widely so the fallout might well be available to the manufacturing community.

In summary, it is safe to conclude that artificial intelligence can play a serious role in the factory of the future. It is already performing in the role of expert in several areas and once sensory data can be interpreted by computer recognition modules, the tools in the factory will come alive and play in harmony to optimize the production of products for mankind.

Part II

Application of CIM

Chapter 3

Application of CIM

Can the benefits of computer-integrated manufacturing be realized in your manufacturing application? One question to ask is "Are the parts to be manufactured in the low to mid-volume range?" Here, the flexibility of the system to manufacture a variety of parts allows the aggregation of all suitable part types to reach the mid-volume range, even though the yearly production volume of some parts may be low.

Parts need not appear similar in shape or geometry, yet they should share certain broad characteristics that will allow them to be grouped into families. Parts should be roughly similar in size and weight, made of compatible materials, and the manufacturing processes should be broadly compatible. Such patterns of similarity usually already exist in conventional plants; in metal-cutting operations, for example, large boxy—or "prismatic"—workpieces seldom if ever get in the way of small turned forgings. Another prerequisite, at least for the current generation of CIM systems, is that nominal precision is required of the final part. Future generations of CIM are expected to produce increasingly precise parts.

Aggregate production suitable for CIM application has typically ranged from yearly rates of 1,000 to 100,000, but the specific processes and process times will decide the profitability of a CIM system. In general (Figure 3–1) at very low-production volumes, stand-alone NC machines (job shop) are still the best choice having the flexibility to adapt to produce a few each of very different part types. At very high production volumes, a very inflexible transfer line will be the most cost-effective alternative.

Figure 3–1 implies that CIM is the only technology to be used in the intermediate production volume range, but this interpretation depends upon how CIM is defined. A restrictive definition of a CIM system will exclude certain technologies which can be applied economically to the

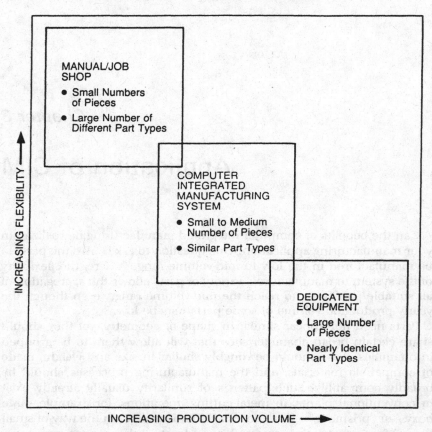

Figure 3-1.   Manufacturing Technology Application areas.

manufacture of medium-volume parts. The potential buyer in conjunction with responsible system suppliers must decide which version of the technology is appropriate in a given situation.

Typically, as with any state-of-the-art technology, the successful acquisition of CIM requires the organization to have an influential "believer" in the technology to "champion" the concept. Ideally, this individual is technically oriented, will head the team and will provide continuity for the project as well as being the primary contact for the CIM supplier. The acquisition of a CIM system requires the understanding and support of the plant's personnel as well as that of top management.

Challenges to Management and the Organization

Compared to stand-alone NC machine tools, robots, and work cells made from them, a CIM system consists of many interconnected components—from the computerized control system that oversees the entire operation with its hardware, software, sensors, actuators, and communication links, to the automated material handling system with its carts, pallet shuttles, and/or robots, to the computerized production planning system. Due to the sparse manning of the system, strict tool-management discipline for a large number of tools, meticulous incoming inspection, and rigorous attention to maintenance is required. In an automated system, machinists should not be fixing problems "on the fly."

All departments, from accounting through quality control and beyond, feel the impact of a CIM installation. Not only must systems and procedures be reorganized, but the attitudes and skills of people in the company have to be altered and upgraded. It is important to recognize this at the beginning.

An appropriate infrastructure is needed. The organization should have a staff of technical and financial planners, and it is beneficial to already have a computer facility staffed with personnel experienced in real-time hardware/software problem solving.

And finally, getting a complex CIM system up and running requires full cooperation between seller and buyer for a considerable period of time. Even with operations personnel working side by side with the vendor during installation, it may take up to six months before the system reaches the designed production volume.

CIM Acquisition

If CIM technology appears to be compatible with your strategic goals, then a more detailed examination of the specific application is warranted, as described in Chapter 4.

Establishing a CIM Task Force

Once the preliminary decision to procure a CIM system is made, the next step is to organize a task force. This group will define goals, develop a schedule, learn the current state of the technology, structure bid invitations, evaluate the resulting proposals, and plan for the system's installation and operation.

The task force should be headed by a high-ranking member of the organization, for example, the vice president for manufacturing. It should include a representative from nearly every department that will be affected, including a technical planner, a financial planner, a manufacturing-operations engineer, a systems engineer, a computer engineer, a product-design engineer, and a maintenance engineer. Since task force members will serve as a primary means of communication between the people who install the system and those who will run it, the members should be motivated and knowledgeable decision makers in their departments.

CIM Implementation Plan

The assembled task force should begin by defining goals and laying out a rough schedule for system acquisition and use. Some plan elements would include the degree of flexibility desired, the minimum acceptable level of reliability (availability), the system's probable location, the means of financing, the desired return on investment, the goal for reduction in inventory, the average machine utilization goal, etc.

A key plan element is the determination of which part family or families may be run on the CIM system as well as its anticipated production rates. The specification for the family of parts will include a list of operations: milling, drilling, tapping, turning, boring, punching, stamping, hardening, etc. It will also include the required maximum workpiece envelope size, the appropriate type of fixturing, and the materials that will be processed.

Further, the task force should identify potential interface problems between the proposed CIM system and the present manufacturing components. Such problems might occur in a variety of areas including accounting, production planning, inventory control, tool management, computer facilities, and labor relations.

The Request-for Proposal

Following some interaction with potential CIM suppliers to determine the current state of the art, and after some team members visit an existing CIM facility (where feasible), the team's next task is to draft a set of specifications. These will form the core of the formal request-for-proposal (RFP) with additional items, such as installation schedule, acceptance test descriptions, training requirements, and computer-simulation requirements supporting the document.

In order to speed up the CIM acquisition process, it would be advisable to present a preliminary copy of the bid invitation to potential vendors for comments. Also, prior to issuing the final version of the RFP, consider a small contract with a CIM vendor or a consultant to do a detailed analysis of the candidate CIM system. Such a study would include a computer simulation of the system's operating performance parameters.

Evaluations

Vendor proposals will cite most of the system costs, and these, of course, should be compared. Other unique costs that the purchaser will add into the calculations include those for specific site preparation.

More difficult to evaluate will be the extent to which each proposed CIM system will meet or exceed the specifications set in the RFP. Computer simulations of each proposed system will help predict productivity. The reliability and accuracy of each CIM, as well as other intangibles, are more difficult to predict and compare. Therefore, the reliability and accuracy experience obtained on similar installations should be examined.

Machining centers (and turning centers) are very flexible in their capabilities, yet they effectively automate production only one part at a time. Dedicated production equipment, such as transfer lines, on the other hand, often sacrifice ease of changeover for repetitive accuracy and speed of throughput for "mass" production.

In the middle lies an opportunity to economically automate midrange production volumes to achieve responsiveness. CIM technology addresses this area of manufacturing. CIM technology represents an opportunity to use computer techniques to boost productivity, by increasing machine use and making a manufacturing plant extremely responsive to changes in demand.

Taken as a whole, this information will provide guidance. Users must provide the expertise, the resources, and the will to make it compatible with their unique situations.

Justifying the Need

Justification Criteria

The primary reason that justifies CIM in a manufacturing environment is economics. CIM and its peripheral devices are capital equipment and are expected to provide capital return, hopefully a profit over its useful lifetime. Among the factors considered are the payback period, net present worth, equivalent annual worth, internal rate of return, and average return on investment.

Justification for CIM projects will, no doubt, follow somewhat established paths, as most manufacturing activities still have routine procedures for evaluating capital expenditure proposals. There will be several differences from the normal investment analysis including the lack of an experience base, an inaccurate projection of capital and expense costs, a lack of experience making projections, and difficulty in quantifying benefits.

Upper management may, in the interest of recognizing the need to gain firsthand experience, relax the normal economic justification requirements. The CIM manager should determine what the project justification criteria will be, and if they differ significantly from other capital investment projects, assure that those responsible for the preparation of financial analysis and capital request authorizations are aware of the variances.

The next step in validating the process is to identify and quantify the savings and negative cost factors. This can be done in a general way. A list of possible benefits and cost factors will be prepared and values will be assigned to each. For the CIM applications, the appropriate factors will be selected and the actual economic data developed. Analyses, approximations, and estimates of value for benefit and penalty factors will be used.

Benefit Factors

Benefits can be both economic and noneconomic. The economic factors are of the greatest importance since they are the ones that can be quantified and later measured and compared. The noneconomic benefits (such as improved quality), usually presented for management's consideration in the final decision process, are a subordinate consideration.

Direct Labor Costs.   For automation that replaces existing operations, one of the significant economic benefits is derived from the displacement of labor, the removal of some or all of the direct labor assigned to the present operation. The value of the direct labor savings has several components including wages, fringe benefits, cost-of-living allowance, vacation pay, health care, and pensions. Other factors affecting labor include state and federal costs, such as worker's compensation, unemployment insurance, social security tax (FICA), and the like. Labor rates used in quantifying benefit factors should consider increases in wages, fringes, and costs over the life of the CIM application.

There are other labor-related economic benefits that are also proportional to the direct labor displaced including shift premiums paid for second and third shift operations, clean-up, and other paid nonproductive time. These costs may range from 5 to 20 percent of the direct labor rate.

Indirect costs of direct labor may be included in a "burden" rate applied to the direct labor wage rates and include protective clothing and personal safety devices (glasses, hard hats, hearing protectors, and so forth), periodic hearing and toxicology tests, workplace lighting, heating, cooling, and other costs associated with ensuring workers' safety and comfort.

Using CIM has other advantages. Poor working conditions such as noise, dust, fumes, heat, dirt, heavy loads, fast pace, or monotony often lead to work stoppages or slowdowns, uncompleted operations, poor workmanship, grievances, absenteeism, or high labor turnover creating higher than normal operating costs. Overtime to make up production losses, rework, and repair will necessitate additional labor and administrative costs. These may amount to 5 to 10 percent of the direct labor costs associated with the operation.

Avoidance of potential costs arising from disability claims will be a benefit of CIM applied in specific operations, although quantification will be difficult. Many diseases and disabilities are being attributed to long-term performance of certain industrial tasks including exposure to industrial chemicals. Using CIM over a long period of time may effectively avoid the future cost of some disability claims.

As with any capital asset, a CIM system can be depreciated and considered a capital benefit factor. The depreciation schedule can extend over the expected life of the system, which may be as long as ten years or more, but an accelerated depreciation schedule may be considered.

There are many conditions to be considered in determining the utilization schedule and its potential productivity increase due to CIM. The upstream or preceding operations must be able to support the CIM center schedule to prevent waiting time. The operations that follow must be able to absorb the increased throughput of the CIM center or processed material must be removed from the operation at such a rate as to preclude waiting time. The overall manufacturing schedule must be designed to absorb additional parts or material produced. Productivity increases on the order of 10 to 35 percent are commonly gained.

CIM work cells. Quality improvement can result from the consistent operation of a CIM installation. The correct selection, alignment, and installation of components during assembly operations can be assured to eliminate the variability and error of human operations from the manufacturing process.

Included in the primary benefits are reductions of scrap, material usage, energy consumption, repair, and rework. Material usage in finishing operations can be reduced by 15 to 20 percent, scrap reductions may average 5 to 15 percent, repair and rework costs may be reduced by as much as 20 percent in some operations. Improvements in quality may amount to 5 percent of the value of the parts produced.

There are three aspects to inventory expense:

1. Inventory value or cost,
2. Length of time in inventory, and
3. Carrying cost of inventory including: cost of money, taxes, storage, handling, obsolescence, damage, and loss.

Quality improvement and increased productivity generated by a CIM system can lead to lower inventory expense. Quality improvement reduces scrap and material usage, and decreases raw material inventory requirements and in-process inventories. Increased productivity shortens lead times, reduces work in process, and finished goods inventories.

The cost of inventory can average 2 to 4 percent per month while the quality improvement and productivity increase may reduce the value of the inventory of the parts involved by 10 percent and length of time in inventory another 10 percent.

Indirect Benefits

CIM, by virtue of its programmability, provides more flexibility to an operation than fixed or special-purpose automation and thereby provides indirect economic benefits. This provides the capability to meet changing demands by varying production rates on products without changing the size of the work force and introducing new products into the manufacturing system more quickly with little change to production facilities.

The enhancement of competitive position has both direct and indirect economic implications. The obvious direct economies are reducing the cost of producing goods and allowing for a profit or pricing advantage over competitors. Indirect economic advantage lies in the inherent flexibility of CIM. In the early stages of CIM it is unlikely that any significant competitive advantage will be gained and the economic value of competitive position, at this stage, is small.

CIM is sometimes justified solely on the basis of management's direction, without regard for economic considerations. Besides the potential waste of capital, time, and resources, if the application ultimately fails, a bad experience may discourage management from further use of CIM technology. Upper management support and encouragement are critical to the success of any CIM project, however, these do not in any way obviate the need to approach CIM in a systematic professional way.

A list of benefit factors and some typical values to be used in the justification analyses are shown in Table 3–1.

Justification Impacts

Economic penalty factors include the initial and continuing costs for the CIM workcell components and peripheral equipment. Initial costs include expenditures involved in implementing CIM for a specific operation, from application engineering through equipment procurement and site preparation, to the actual installation and operation, and are basically all of the one-time costs. The costs of the initial survey, qualification, prioritization, application selection, and justification are not often included in the initial costs of any particular CIM installation.

The cost of CIM includes machine tools, robots, material handling systems, computer hardware and software, and interfaces as well as accessories, terminals, spare parts complement, special tools, and so forth. Also included in the cost are hazard barriers, intrusion detection devices, and other safety equipment associated with the installation. Interlock hardware between the CIM setup and other machines or equipment in the area would also be charged against this item.

Tooling costs must include any special purpose devices. Among the items that are considered in the cost of tooling are the end effectors for each robot, any adaptor plates or tool mounting devices, parts feeders, orienters, special power tools (such as self-feeding screw drivers), fixtures and pallets for positioning and retaining parts, and assembly fixtures. Not considered as tooling are machine tools, conveyors, and other items of factory equipment that could be used in a nonautomated operation. Because tooling may be task-specific, it may be expensed rather than capitalized, or it may be written off on a different depreciation schedule than the robot, conveyor, machine tool, or similar capital assets.

Facilities include conveyors, machine tools, presses, molding machines, other production equipment purchased as part of the application, and containers, pallets, stock racks, and dunnage.

Modifications to machinery and facilities to accommodate a CIM installation will be considered as a revision or rearrangement expense.

Table 3-1. CIM Benefit Factors and Values

Factor	Typical Value
Direct Labor	
Wages	Fully burdened, long-term average
Fringe benefits	hourly rate
medical and life insurance	
Cost of living allowance (COLA)	
Vacation pay	
Retirement pensions	
State and federal costs	
Workers compensation	
Unemployment insurance	
FICA (Social Security)	
Related Direct Labor Costs	5 to 20% of direct labor rate
Relief allowance	
Shift premium	
Wash-up and other nonproductive time	
Related Indirect Labor Costs	
Supplies and expenses	Included in burden on hourly wage
Proective clothing and safety equipment	rate
Hearing and toxicology tests	
Lighting, heating, cooling, etc.	
Undesirable operations	5 to 10% of hourly rate on high turnover jobs
Absenteeism allowance	Up to 5% of direct labor work force in high turnover industries
Safety	Differential between cost of other
OSHA and other regulations	compliance means and robot
Reduced workers' compensation	
Other insurance expenses	
Investment Tax Credit	Up to 10% of capital investment
Depreciation	Based on life of robot application
Increased Productivity	Up to 10% greater throughput
Higher throughput	
Higher machine and facility utilization	
Improved Quality	
Reductions in:	Up to 5% of value of parts produced
Scrap	
Material usage	
Energy consumption	
Repair/rework	
Reduced warranty and field maintenance/repair	Not quantified
Increased customer satisfaction	Not quantified
Reduced Inventory	
Value of raw material, in-process, and finished goods inventory	Up to 10% reduction
Time in inventory	Up to 10% reduction
Cost of inventory	Average 2% per month
Potential Cost of Disability	Not quantified
Advancement of Technology	Not quantified
Increased Flexibility	Not quantified
Competitive Position	Not quantified
Management Direction	Not quantified

Items include relocation or repositioning of machines or ancillary equipment, rerouting of conveyors, modification of machine tool splash guards, blow-off, coolant/cutting fluid systems, chip handling systems, changes to operator interface panels on machines, and so on. Providing utilities such as compressed air, electrical power, or cooling water, if not otherwise required to support the process or operation, should also be included. Modifications to floors, special machine foundations, anchor bolts, bases or platforms, and similar changes are also covered here.

Other expenses of the installation include supporting equipment, routing and terminating interlock cables, power, air, and water lines, erecting hazard barriers, and installing other safety and communication systems.

Training operators, maintenance personnel, and supervisory personnel is ongoing because of transfers and attrition. Training is normally a continuing expense, however, initial training programs for CIM system operators and for personnel responsible for system maintenance, normally provided in advance of the installation, are considered an initial cost. In addition, the cost of retraining employees displaced by the CIM system to qualify them for job reassignment and/or the cost of transferring displaced personnel to other jobs or locations is also included in the initial cost.

Revising the product to facilitate a CIM application may also be required. The cost of such product changes, which may include engineering, design, and tooling, are chargeable to the initial cost. Cost differentials between old and new parts, other than manufacturing costs, should be included as either a continuing cost penalty or as a savings benefit.

One item of initial cost generally overlooked is the cost of packaging, dunnage, and containers used to store, deliver, and orient parts in CIM workstations.

Once the CIM system is installed, there will usually be a cost penalty incurred during the programming, start-up, and debugging of the application. Interferences with production, stockpiling of parts, overtime to make up for production losses, damaged or scrapped parts, and repairs to equipment damaged by program errors are a few of the sources of extraordinary expense normally experienced during the launching of a system.

Continuing Costs

There are a number of ongoing expenses associated with a CIM system. These can be classified either as operating expenses or as overhead.

- CIM will require continuing maintenance. Where robots are installed to replace manual operations, there will be an incremental increase in the maintenance staff. The experience of major users indicates that servo-controlled robots and their associated tooling and related equipment require an average of around thirty minutes of maintenance labor per operating shift.
- Cost of maintenance supplies and spare parts for CIM components varies widely. Fixed-stop robots use relatively inexpensive components compared with servo-controlled robots. Hydraulic robots tend to leak and require replenishment of hydraulic fluid, which is not necessary with electric robots.
- Another expense is the operating energy cost.
- There are also overhead items that are part of the continuing cost penalties. These include: cost of capital, training, packaging, and taxes and insurance.
- Investing in CIM is capital intensive. Most cost analysis methods consider the cost of capital, which reflects the interest rate paid for the funds to purchase and install the system, the cost of leasing the equipment, or the interest that the funds would generate if invested instead of spent.
- As mentioned previously, training is both an initial investment and a continuing cost. After the first training program, which may be included in the price of the CIM, follow-up training can cost a minimum of $2,000 per trainee. Typically, replacement and refresher training programs may involve two to three people a year.
- The use of special packaging for CIM operations is increasing. In some cases, especially for handling parts produced in-house, this special packaging may be designed to be reused. For parts produced outside the plant, the packaging is often not reusable. The original complement of such material is included in the initial cost of the system; however, replacement or repair of special packaging, dunnage, and the like is considered an overhead item and part of the continuing costs.
- Two elements of taxes that are part of the CIM overhead are the property tax levied against the capital assets (the CIM work-cell systems) and the income tax assessed against the profits gained from the operation. There may also be incremental increases in other costs of business such as insurance on plant and equipment and so forth.

Table 3–2 lists some of the penalty factors and their initial costs for an industrial robot involved as part of a CIM facility.

Table 3-2.   Penalty Factors—Initial Costs

FACTOR	TYPICAL VALUE
Robot	
Manipulator	
Safety equipment	
Interlocks	
Tooling	
End effector	
Part feeders, orienters	
Fixtures, pallets	
Facilities	
Conveyors	
Containers	
Facility rearrangement	
Relocation of machines	
Modifiation of machines	
Utilities	
Foundations	See below
Installation	Total value =
Robot and ancillary equipment	1.75 to 2.65
End effector	times cost
Safety equipment	of the robot
Application Engineering	
Process Changes	
Engineering and documentation	
Product Changes	
Engineering, design, tooling	
Packaging	
Containers and dunnage	
Training and transfers	
Operators; maintenance personnel, supervisors	
Overtime	
Damage and Scrap	
Maintenance Labor	
Fixed-stop robot	15 min. per operating shift
Servo-controlled robot	45 min. per operating shift
Maintenance Supplies, Spare Parts	
Fixed-stop robot	$0.60 per operating shift
Large electric servo-controlled	$3.50 per operating shift
Large hydraulic servo-controlled	$4.50 per operating shift
Energy	
Pneumatic or small electric	$0.50 per operating shift
Large electric	$1.00 per operating shift
Large hydraulic	$3.00 per operating shift
Product Changes	Cost differential
Training	
Replacement personnel and refresher courses	About $2,000 per trainee

Summary

The justification process is created after selecting potential applications in order to support further CIM system implementation. It is done in several steps, beginning with determining justification criteria, followed by identifying and quantifying potential benefit and penalty cost factors, and applying these factors to the selected applications. The major benefit factors are economic since they are the only ones that can be quantified and then measured and compared for alternate applications and conditions. Noneconomic benefits are presented also for management's consideration in the final decision (selection) process. Penalty factors are primarily economic and include both initial and continuing costs for the CIM system and peripheral equipment.

The CIM Implementation Process

Presurvey Activities

The application of a computer-integrated manufacturing system requires careful identification and consideration of all future operations and the impact of the new technology on manufacturing operations. Implementation begins with an initial survey of the entire manufacturing facility which will generate opportunities (operations) for which CIM might be applied.

The first of these presurvey activities is to become familiar with the CIM technology and products. One source of product data is the manufacturers' and distributors' published literature describing their products. The engineering department should establish and maintain a current product reference file and others involved in the implementation process should attend national or regional CIM conferences and exhibitions to maintain a current awareness. Participation in a local chapter of the Society of Manufacturing Engineers (SME) is another way to maintain awareness and to explore issues of common interest with others in similar situations. There are also videotapes available through SME covering basic technology, case studies for other CIM installations, descriptions of system interfaces involved in CIM, and a variety of specialized topics.

Another presurvey activity establishes upper management's position relative to CIM. Firm support and policy statements are required for successful implementation. The implementing manager needs a clear policy relative to three issues: economic and other justifications for CIM, long-term commitment to automation and modernization, and disposition of workers displaced by CIM.

Upper management must respond to questions raised about the impact of CIM on the work force. Media focus has been on the potential large-scale loss of jobs; predictions range from one to three million jobs lost to automation by the middle of the 1990s and possibly as many as seven million by the year 2000. Predictions of job creation are of similar magnitude, ranging from 800,000 to 1.5 million new opportunities for CIM technicians and engineers by 1990. Thus, upper managers may be concerned about the public's view of them as creators of unemployment and, at the same time, may worry about the availability of skilled people to keep their equipment running.

Implementation managers should make available to decision managers sufficient data to resolve the labor issues. In this way, consistent corporate support and policy can be established. Upper management's support will also eliminate much of the resistance to CIM often found in middle-management ranks.

The last presurvey activity is organizing the automation survey team. Successful implementation of CIM is a team effort and should include representatives of all of the affected organizations in the corporation. Involvement will usually encourage support. The team should be designed so that its in-plant activities are not disruptive and so that it can reach consensus on recommendations and actions. One approach is to make up a team with representatives of all the organizations involved and then set up smaller task forces for specific implementations. The team might include members from manufacturing (tool or process) engineering, industrial engineering, production management, plant (facilities) engineering, quality control employee (labor) relations (personnel), cost accounting (controller's office), product engineering, and direct (production) labor. Similarly, the task force performing the actual in-plant survey might only include manufacturing engineering, industrial engineering, and production management. Of course, they should be provided with copies of policies, procedures, and schedules.

After the survey team is established, a coordination meeting should be organized to introduce the various team members to each other; to review the objectives of the team; to discuss process and procedures by which the team will operate; to establish responsibilities, roles, and organization of the various subgroups or task forces; and to review the overall survey schedules.

Initial Survey

After the coordination meeting is held, the actual agreed-upon process can proceed. The subgroup force doing the initial survey should begin in-plant activities. Unless the plant is relatively small, the survey should be done in parts, one department at a time. The production supervisor for each area should be notified in advance prior to the appearance of a group of people looking around, asking questions, and taking notes. The team should ask questions to clarify their impressions and answer questions put to them by supervisors and workers, although a detailed analysis of operations will not be done at this time.

Of special concern will be:

- Equipment in place
- Production requirements
- Manufacturing costs
- Labor turnover
- Absenteeism
- Quality
- Worker complaints
- Worker safety
- Severe working conditions
- Accidents and injuries

The employee or labor relations representative on the automation team can be a source of information about turnover, absenteeism, and worker complaints. The production labor representative or plant engineering representative can identify severe working conditions. Other team members can provide information about safety, quality, and cost issues, and similar data can be obtained relative to accidents and injuries.

There are several reasons for concentrating on these areas: (1) removing workers from these operations will meet with little resistance, (2) upper management may allow special consideration in justifying automation in these operations, and (3) potential savings may be higher than for other average operation. At the same time, those very conditions that create problems for human workers may make automation more difficult than usual, and such operations must be approached with the same degrees of caution and objectivity as any other in the plant.

As previously noted, the objective of the initial survey is to develop a list of potential applications for automation. During this step in the process, prejudgment should be avoided (both negative and positive) and a detailed analysis of each potential application should not be performed. Rather, the team should look for tasks that meet certain broad criteria:

- An operation under consideration must be one that is physically possible for CIM implementation.

- An operation under consideration must be one that does not require judgment.
- An operation under consideration must conform to justification guidelines for the use of CIM.

The team should proceed systematically through the area, considering each operation for a potential application. Many operations will clearly not meet the broad criteria. Questionable operations should be placed on the list, the qualification and prioritization steps that follow will ultimately eliminate those that are not feasible.

With regard to justification, the economic attractiveness of a potential CIM application, as measured by return on capital or by payback period, is usually of primary importance. When incorporating automation, a major source of savings is in reducing labor cost that results from displacement of workers. A rough estimate of these savings can be made based upon the potential labor that could be displaced. With predicted savings and an estimate of the installation costs, a rate of return or a payback period could be calculated. For purposes of the initial survey, however, a simple rule of thumb may suffice.

The aggregate list will include the identification of each potential application. The area of the plant where the operation is observed should also be recorded. The number of production shifts should be noted. Brief comments about the operation should be included, with reasons for its selection. In the case of questionable operations, the areas of concern should be recorded.

Nonmanufacturing areas should be reviewed as well as production operations; high potential applications may be found in receiving and warehousing and in packing and shipping, as well as in manufacturing. The initial survey should be completed before any analysis or qualification is undertaken.

Qualifying CIM Applications

Qualification will be an iterative process, the potential applications should be reviewed at least twice. The qualification step involves the entire CIM survey team. Each team member should be provided with sufficient time to familiarize themselves with the proposed applications list.

The qualification review should be conducted in an orderly manner, using the developed list as the agenda. Each potential application should be described by a member of the initial survey task force. All team members will comment or contribute whatever specific information exists about the operation under consideration. Unanimous agreement on the qualification or disqualification of each potential application is desirable but not likely to be attained; however, a consensus agreement should be reached.

Application Reviews

The application review involves addressing each potential operation.

- Complexity of the operation
- Degree of disorder
- Production rate
- Production volume
- Justification
- Long-term potential
- Acceptance

A second review reconsiders each of the applications that remain after the first scrub, while visiting each operation on the factory floor with the entire team, again considering the seven factors. By reviewing each operation on the refined list, we can eliminate those where CIM should not be used.

There is sometimes a tendency to misapply CIM. Careful consideration of the seven factors and the simple rules for each should reduce this likelihood.

CIM Implementation

An industrial CIM system can represent a significant investment in capital and in effort. Economic justification must therefore be carefully considered. A balance sheet of positives and negatives would contain benefits such as increased productivity, reduced scrap losses or rework costs, labor cost reduction, improved quality, improvement in working conditions, avoiding human exposure to hazardous, unhealthy, or unpleasant environments, and reduction of indirect costs. Typical nonbenefits would include capital investment, facility, tooling, and rearrangement costs, operating and maintenance expenses, special tools, test equipment and spare parts, cost of downtime or backup expense, and the costs of integrating CIM into the production environment (training and so forth).

Estimates of costs and savings should be made to determine if reasonable economic benefits can be expected. The savings can be roughly estimated by multiplying the number of people displaced per shift by the number of production shifts per day by the fully burdened annual wage rate. The costs can be roughly estimated by multiplying the basic cost of the CIM system planned for the operation by 2.6. Dividing the estimated cost by the estimated average yearly savings gives an approximation of the payback period in years. Generally, a payback period of two and a half to three years is acceptable; for qualification purposes, a two-year payback is recommended.

Neither management direction nor emotion are substitutes for economic justification and will not support the application of CIM. Safety or working conditions may override economics, however, these are usually exceptional circumstances.

The number of potential applications and their expected duration must be taken into account. Because of its flexibility, CIM can usually be used for other applications if the original operation is discontinued. Since the system's useful life may be as long as ten years, several reassignments may be made in the future.

CIM interfaces require special knowledge and skills to program, operate, and maintain the system. An inventory of spare parts should be kept on hand. Auxiliary equipment for programming, program recording, maintenance, or repair may be required. Personnel training, an inventory of spare parts, special tools, test equipment, and the like may represent a sizeable investment.

Not everyone welcomes automation. Production workers are concerned with the possible loss of jobs, factory management is concerned with the possible loss of production, production supervisors are concerned with managing new technology and motivating machines, maintenance personnel are concerned with learning new skills to understand the new technology, and company management is concerned with effects on costs and profit. Reassignment of workers displaced by automation can be disruptive. Training personnel to program and maintain the system can be very expensive, upset maintenance schedules and personnel assignments, and new skills may even have to be developed. The installation and start-up can interrupt production schedules, as can occasional breakdowns of the equipment.

Chapter 4

Planning for Computer-Integrated Manufacturing Implementation

Can Your Organization Benefit from and Support CIM?

Is computer-integrated manufacturing system technology right for your organization? To answer this question, it will be necessary to examine your future production needs as well as the experience and capabilities of your organization.

If you are now producing, or expect to produce, parts requiring similar manufacturing operations with aggregate volumes in the low to mid-volume range, then CIM may be applicable. These parts could be small quantity orders that are placed a number of times every year, or medium quantity orders which are divided into batches to satisfy the customer's usage pattern or spare parts, that is, neither mass production nor one-of-a-kind items.

Parts that require processing on many different machines, frequent refixturing on the same machine, approximately the same machining cube, and modest tolerances are excellent CIM candidates.

Organizational experience with NC machines, and preferably CNC systems, is important. It will have allowed the work force to develop some of the electronic and computer skills needed to operate a fully integrated CIM system. Other needed skills include site planning, production planning, process planning, part programming, quality control, tool management and maintenance, and acceptance of technological change.

An adequate planning horizon is also required. The lead time for a CIM system can approach two years from the date an acquisition decision is made. Quite often it will require another 12 months after installation to obtain full production. If the CIM facility is going to produce existing parts, the system can be brought up to full production gradually, since the production capability already exists and time is available to correct unexpected problems. When CIM is introduced, adequate time must be allowed for installation and debugging. Rushing the CIM vendor during the design and installation phases often results in a longer shakedown period.

If your production requirements and organization satisfies these criteria, then a CIM implementation will most likely be beneficial.

Defining Manufacturing Goals

One of the most important steps in planning for CIM is to define, as clearly as possible, how the system is to satisfy your present and future manufacturing needs. This will influence the specification to the CIM vendor and will prevent the system from being under-designed or over-designed. Typical CIM manufacturing goals include:

- Improve quality of parts
- Reduction in part manufacturing costs;
- Less skilled labor required;
- Reduction in part production lead-time;
- Reduction in work-in-process inventory;
- Flexibility to produce spare parts or change product mix as market requirements demand;
- Staying at the forefront of technology.

Steps for the Implementation of a CIM

Figure 4–1 through Figure 4–8 summarize the steps required for CIM implementation. Each implementation step in the text has a flowchart to highlight the step in relation to the implementation sequence. With each flowchart there is a list of the tasks to be accomplished and the data required for that implementation step.

Prioritization and Selection

The qualification process will eliminate a number of potential applications. Those that remain should be operations that qualify as technically

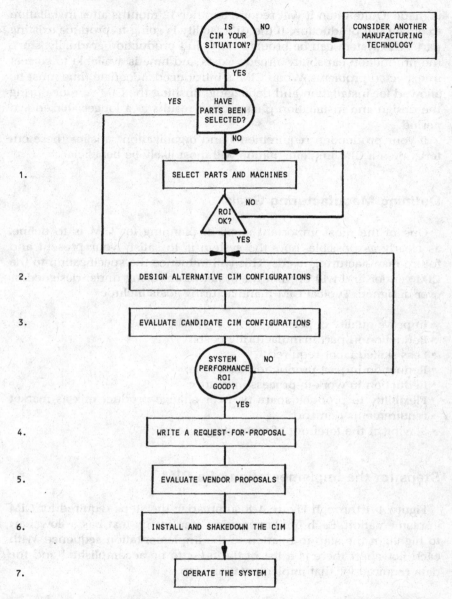

Figure 4-1. Decision Flowchart for the Acquisition of a Computer-Integrated Manufacturing System (CIM)

- Preselect parts and machines having CIM-compatible attributes from available candidates.
- Calculate current production cost of each part.
- Estimate CIM costs for each part.
- Use either manual selection methods or a computer software package to select the most economically beneficial parts and machines.
- Perform investment analysis to determine if CIM is an economic alternative.

Figure 4-2. Step 1—Select Parts and Machines

1. Estimate the Work Content of the Selected Parts
   - Develop CIM fixturing concepts for the selected parts, minimizing the number of fixturings.
   - Process plan each part in detail, constrained by the limited tool capacity of CIM and the effects of using different machines on overall accuracy and cycle time.
   - Determine the appropriate machinability data for each material, for each class of operation.
   - Estimate production requirements for each part.
   - Calculate part cycle times and tool usage.
2. Design Several Equipment Configurations
   - Choose specific vendors' equipment in each machine class; adjust to reflect company preferences (toward horizontal rather than vertical machining centers, for example).
   - Estimate the minimum number of machines for each machine class.
   - Modify this number of machines to account for shop and system efficiency, limited tool storage capacities, and desires for machine redundancy.
   - Add a material handling system (MHS) and other desired nonmachining processes, such as an inspection machine, to complete the configuration.
   - Layout the equipment and material handling system.
   - Develop alternative design configurations from the original design.

Figure 4-3. Step 2—Design Alternative CIM Configurations

- Simulate the operation of each configuration based on predetermined scheduling, batching, and balancing rules to provide performance measures for each configuration.
- Improve the configuration designs until each provides satisfactory performance measures or is rejected.
- Perform a detailed investment analysis of each configuration.
- Examine and evaluate intangibles, such as flexibility, accuracy, etc.
- Choose the configuration which best satisfies the investment and intangible analyses.

Figure 4-4. Step 3—Evaluate Candidate CIM Configurations

- Write an RFP that conveys your findings and desires for a CIM system.
- Avoid overspecification; allow the CIM vendors to be creative and competitive in designing a system for your applications.

Figure 4-5. Step 4—Write a Request-for-Proposal (RFP)

- Verify and evaluate vendor proposals using simulation and economic analysis.
- Verify the financial stability of the vendor.
- Evaluate the degree of success of each proposal in satisfying your nonquantifiable requirements.
- Choose the proposal which best satisfies your company's need.
- Work with the vendor to develop detailed specifications and prices.
- Place an order.

Figure 4-6. Step 5—Evaluate Vendor Proposals

- Select and educate personnel to operate and maintain the CIM system.
- Assess the quality control and production control departments' roles in the successful implementation and operation of the CIM and develop or augment policies to assure success.
- Develop a preventative maintenance plan and spare parts lists for the CIM system.
- Prepare the site.
- Assist vendor with installation and shakedown.
- Perform your acceptance tests.

Figure 4-7. Step 6—Prepare for, Install, and Shakedown the CIM System

- Schedule parts.
- Batch production if necessary.
- Allocate parts and tools to machines.
- Balance machine loads.
- Use a decision support system to optimize daily operations in the face of machine failure and changing part requirements.

Figure 4-8. Step 7—Operate the System

and economically feasible. Once the priorities of potential applications are established, the sequence in which they may be implemented can be determined. Assuming that CIM is new to your manufacturing operations, the selection of the applications must be done with care to minimize the chance of failure with the attendant CIM interface developed as part of the initial setup.

Prioritization is a follow-on qualification and should be undertaken as soon as possible after qualification. Prioritizing should be as objective as possible, although some subjective factors will have to be considered.

The prioritizing of operations and subsequent selection of an initial CIM application can be facilitated by the use of an operations scoring system. The elements of the scoring system might include:

- Complexity of the task;
- Complexity of tooling, part orienters, feeders, fixtures, and the like;
- Changes required to facilities and related equipment;

• Changes required to produce and/or process;
• Frequency of changeovers, if any;
• Impact on related operations;
• Impact on work force;
• Cost and savings potential;
• Anticipated duration of the operation.

For each of the elements involved, a set of measures and a score range is established, with the more important elements having a greater range of points than the less important.

Selection of the elements of the scoring system may be somewhat subjective. The uniform application of the scoring system will also reduce the final subjectivity of the rankings.

The collection of data is best done by input from individual members of the survey team, in their areas of knowledge. The industrial engineering representative should provide data on labor savings, cycle times, and the like; manufacturing engineering should give input on ease of orienting parts, complexity of tools and fixtures, and so on; industrial relations should provide input on environment and labor turnover; and product engineering should comment on potential product changes. Hard data will not be available for all elements of the scoring system; in such cases, the team members should provide a best estimate and the data should be so noted.

After the data have been collected, the entire survey team should reconvene to establish scores. After a score has been assigned to each element, an overall score should be computed for the operation and noted on the sheet. Once all qualified operations have been scored, the final phase of the prioritizing process can begin.

The establishment of implementation priorities for the qualified operations should be based largely on the scores established for each. The operation with the highest score should be the prime candidate for the first application. There are, however, some less objective considerations to be factored into the process at this point.

The Selection Process

The selection of the first CIM application must minimize the chance of failure. Failure with the first installation can severely curtail, if not eliminate, any further efforts. After all qualified operations are reviewed during the prioritizing process, it is imperative to again review, on-site, the top candidate operations before making the final selection to ascertain that they are indeed technically feasible and justifiable.

The final selection should be supported by a strong consensus of the survey team and should reflect a high level of confidence in its success. It is advisable to also select one or two alternatives to the recommended first application, in the event of unforeseen circumstances that might arise during the next steps of the implementation process.

Reporting to Management

The final output of the prioritizing and selection process should be a status report to upper management that summarizes the activities to date of the CIM survey team and includes:

* Lists of the members of the survey team and of the task force or sub-group that performed the initial survey. These should be appendixes to the report.
* The list of potential applications generated in the initial survey.
* A prioritized list of qualified operations that have CIM potential.
* The recommended first application and the alternative applications selected.
* A discussion of the rationale supporting this recommendation.

After the report has been submitted to management, a review meeting should be scheduled for the CIM survey team and key management personnel from production, finance, and planning, as well as the top management level. The meeting should present a summary of activities and recommendations. The primary objective is to obtain the concurrence of key managers and of upper management. A secondary objective is to expose management to the potential of CIM, as represented by the list of qualified operations. This meeting will be an important indicator of the strength of continuing management support for the program.

Economic Analysis

The Implementation Decision Process

In reaching a meaningful decision in the implementation of capital equipment, there are questions that must be answered related to economic justification issues, the need for the equipment, the demand for and/or apparent lack of capacity that requires the equipment, the nature and type of equipment under consideration, and its compatibility with the strategic manufacturing requirements for the product. The justification must consider all these factors.

It is important to ensure that nothing has been overlooked since the economic analysis will be a strong input to management for the commitment of resources.

Justification Analyses

At this time, the objective is the analysis of the first recommended application and each of the alternates selected to affirm that they can be justified and to establish the basis for justification.

The analysis team should set up the economic benefits and penalties to be used in the justification analyses, examining the typical values presented to determine those that are appropriate for the situation. Where the specific situation, or where economics have not been presented, the group should come up with its own set of numbers. For example, man-hours of direct and indirect labor can be converted to dollars by applying the current labor rates. Overhead and burden rates can be established. The number of production shifts and workdays per year can be determined.

For the applications under consideration, potential direct labor savings (man-hours per shift) should be estimated. Other sources of savings, if significant, should also be noted. Penalties or costs should also be estimated. CIM directories should be reviewed or a sample group of manufacturers or suppliers should be contacted to develop a range of costs for estimating purposes. Any significant continuing cost factors should be noted and quantified. Any important noneconomic or unquantified benefit and penalty factors should also be recognized.

Once data have been gathered, simple justification analyses can be performed to provide comparisons between the applications being considered. At this point, a formal cost analysis of each potential application is not necessary. However, some measure of economic performance must be applied so that relative merits of the applications can be compared. The economic test is to determine payback period for each application. Payback determinations can be facilitated by use of a simple cash flow matrix.

Typically, a payback period of two to three years is acceptable for capital projects such as CIM installations.

Sensitivity Analysis

Often, as part of an economic analysis, it is desirable to explore a variety of tactical issues related to the implementation of CIM. These issues include:

- Uncertainty in product demand;
- Single shift versus multiple shift operation;
- Capacity availability in related upstream and downstream operations;
- Effects of an increase or reduction in capital investment;
- Extent to which reduced cycle time and/or increased availability could affect its affordable price.

While irresponsible spending is not advocated, it is prudent to provide the system with as much capability (in both automation and peripheral equipment) as available capital will allow. Knowledge of how much investment a particular application will support while still meeting established economic justification criteria could be invaluable to the manufacturing manager.

To assist the manufacturing manager and others, a simple computer program can be designed in BASIC to be capable of quickly answering what-if types of questions which may arise during the economic analysis. It can be accomplished through an after-tax present worth assessment of cash flow differences between the manual operation and the automated operation (net present worth and equivalent annual worth). Technology Research Corporation's *CIMJustification* software, an IBM PC computer analysis tool for CIM economic analysis, is available to provide such a framework. The software is available for $99.95 from Technology Research Corporation, Springfield Professional Park, 8328-A Traford Lane, Springfield, Virginia 22152, (703) 451-8830.

Inputs to the program include

- Labor costs
- Operating costs
- Planned production schedules
- Indirect labor savings or cost
- Maintenance savings or cost
- Other savings or cost
- After-tax minimum attractive rate of return
- Income tax rate
- Inflation rate
- Depreciation rate
- Salvage value

Measuring Economic Performance

There are four measures used to justify capital expenditures; payback period, net present worth, equivalent annual worth, and internal rate of return. Some of these measures are interrelated and in comparing several projects to determine which one has the best economic performance potential, there should be agreement between the results obtained by these various methods.

Payback period is widely used for determining the economic worth of a project. For payback period calculations the cash flow matrix should include all major cost and revenue elements. The cost elements in the matrix should include, as separate items:

- Initial investment for the system, including tooling, facilities, and equipment revisions;
- Annual operating cost for the system, including maintenance labor, maintenance supplies, spare parts, and energy;
- Annual overhead costs, including taxes and insurance;
- Training.

The revenue elements in the matrix should include, as separate items:

- Annual labor savings, including wages, fringes, benefits, and direct labor-related savings;
- Related annual labor-related indirect savings, including undesirable operations, absenteeism allowance, safety, disability, and so forth;
- Other savings, including material, productivity increase, quality improvement, inventory reduction, and the like;
- Such items as depreciation, investment tax credit, and salvage value.

The common convention is to assume that the cash flows occur at the end of the year, that cash receipts are assigned positive values, and that cash disbursements are assigned negative values. Using these assumptions, a matrix can be developed and the cumulative net cash flows examined. The year in which the first non-negative cumulative net cash flow occurs is the payback period.

As it is usually applied, the payback period calculation does not consider the time value of money. Funds invested in equipment lose the opportunity to earn money elsewhere, for example, in bank deposits. The annual amount of interest or other investment income thus given up is considered to be the minimum amount of revenue or savings that the project must earn to be economically feasible. Typically, this minimum acceptable rate of return, or MARR, is set by company policy.

Net present worth (NPW) of a project is defined as the single present value equivalent (at time = 0) of all present and future cash flows, discounted at the established minimum acceptable rate of return. It is determined from the relationship:

$$NPW = -I + \frac{F1}{(1 + i)} + \frac{F2}{(1 + i)^2} + \frac{F3}{(1 + i)^3} + \ldots + \frac{Fn}{(1 + i)^n}$$

where:

NPW = Net Present Worth, dollars. In general, NPW denotes the present value of a single sum of money, when the present occurs in time assumed to be zero. NPW is also used to represent the single-sum equivalent of a series of future cash flows.

I = the present value of the initial investment in dollars.

F = a future cash flow in dollars; also the future value of NPW at some point in time. In this discussion, F occurs at the end of a year.

i = interest rate (minimum attractive rate of return, MARR), percent expressed as a decimal.

n = total number of interest periods (years in this discussion); that is, the expected life of the CIM system.

In the calculation of net present worth, it is first necessary to develop a cash flow matrix, as in the payback period determination. From this matrix, the cash flow values ($F1$, $F2$, . . . , Fn) are selected and entered in the NPW equation.

An answer with a positive value indicates that the net revenues for the project exceed the minimum acceptable rate of return (MARR) on the investment. If NPW turns out to be zero, it indicates that the project's net revenues exactly meet the minimum required. If NPW turns out negative, the return would be less than the MARR and the project should be rejected. With this method, all projects having a calculated net present worth greater than zero satisfy the established MARR. If one project is to be selected from among many with positive-valued net present worth, the one with the largest NPW should be chosen.

The equivalent annual worth (EAW) method is based on converting all project revenues and costs to an equivalent cash flow series with equal end-of-year payments. If the EAW is defined to be the magnitude of these equal payments, the net present worth of the equal payments will equal the net present worth of the original project revenues and costs for a given minimum acceptable rate of return. The relationship is described as follows:

$$NPW = \frac{EAW}{(1 + i)} + \frac{EAW}{(1 + i)^2} + \frac{EAW}{(1 + i)^3} + \ldots + \frac{EAW}{(1 + i)^n}$$

$$= EAW \frac{(1 + i)^n - 1}{i(1 + i)^n}$$

where:

i = the minimum acceptable rate of return (MARR).

n = the expected life of the system in years.

It is first necessary to calculate the present net worth, NPW, as described in the previous section and then calculate the equivalent annual worth, EAW.

Internal rate of return (IRR) is the rate at which all discounted revenues recover all discounted costs for a project. It is, therefore, the interest rate (i) for which the net present worth (NPW) is equal to zero.

With the internal rate of return method it is easy to compare the IRR with the MARR to see how well the project should do against the established minimum performance expected.

Reporting Economic Performance

The economic measures should be applied to the first application recommended and each of the chosen alternatives. Although any one of these might be sufficient for comparing the relative merits of the applications, it is recommended that each of these measures be calculated. The cash flow matrix is necessary to determine the payback period and to compute net present worth. Payback is a reasonable go or no-go indicator, but it does not provide insight into long-term potential because it ignores cash flows beyond the payback period.

Once the NPW has been calculated, determining equivalent annual worth requires only one additional computation. The annualized value is consistent with accepted accounting practices and may be more acceptable than the other methods presented.

Of all the economic measures described, the internal rate of return is most cumbersome to calculate. However, the results may be most easily understood by upper management, as it is simple and quick to compare the projected IRR with the established minimum (MARR). Thus, it is recommended that these calculations also be completed for each application.

Upon completion of the economic analyses, a summary of the results should be prepared and submitted to top management, the same group that was involved in the selection review. A firm recommendation from the CIM survey team regarding the ensuing course of action—proceed with implementation of the first application, substitute an alternate application, or a similar recommendation—should accompany the report. Backup data, including the results of any sensitivity analyses performed, calculations of economic measures, and the like should be either included as appendixes or noted as available upon request.

Since the analysis will involve comparisons between CIM systems as well as variations of each system, it would be most wise to use a prepared analysis software tool like *CIMJustification*.

Chapter 5

CIM Design Analysis

The Buyer's Initiative

Machine tool buyers traditionally have assumed a very straightforward, and sometimes passive, role in procuring equipment. Typically, the buyer decided what machine was best suited to the parts to be produced, and bought it. Alternatively, he would send part blueprints to one or two selected vendors. The vendor would then match the production requirements of those parts to the capabilities of one of his machines and prepare a quote. The buyer would review this quote and either purchase the machine or look for other vendors.

Such an approach, when applied to procuring complex systems, has met with problems. Many buyers have not understood the large number of system parameters under their control, nor have they understood the strategic and operational ramifications of one specification as opposed to another.

To interface effectively with CIM vendors, the buyer should perform a preliminary design and evaluation of a CIM system as though he were going to build and install it. This enables the buyer's staff to understand all the issues involved and results in a Request for Proposal (RFP) that correctly conveys the requirements to the vendor. It also enables the staff to cooperate knowledgeably with the vendor during system shakedown and to subsequently operate the system at maximum utilization. There are three steps in this process:

1. Select a set of intended parts and machines.
2. Design a number of equipment configurations.
3. Evaluate the design and its variations on technical and economic grounds.

Parts and Machine Selection

This section discusses the most basic issue in the entire CIM design and evaluation process: which parts can be matched to available CIM machines to maximize the cost savings compared with alternative production methods. While many qualitative factors also go into this decision, the driving concern for selecting a suitable combination of parts and machines is still usually economic.

Selection Methodology

Part machine selection based on economics can be performed either manually or via computer, as described in Chapter 4. Manual methods work well for situations where there are less than approximately 40 candidate parts and a small number of CIM-type machines to be considered (see Figure 5–1).

When both parts and machines are to be selected from a larger number of candidates, however, manual methods become cumbersome and time consuming.

Both the manual and the computerized methods choose parts based on the same concept, that of relative production cost savings. The present cost of producing each candidate part, either in-house or purchasing it from a vendor, is calculated first. The costs to produce each of those parts in a CIM system are estimated next. Then the parts with the largest savings are chosen to fill the capacity of the selected machines.

The procedures for both the manual and the computerized part selection method are delineated in Figure 5–2 and described in the following sections.

The two alternative procedures are described below:

Manual Analysis

1. Determine initial guidelines.
2. Preselect from the candidate parts and machines those which have CIM-compatible attributes and establish part families.
3. Collect data on candidate parts including process routine, process time, and current manufacturing cost data.
4. Estimate the CIM costs of each part.
5. Choose specific vendors' machines in each potential CIM machine class and obtain prices.
6. Select parts based on relative savings between current and CIM production methods to load a chosen set of machines.
7. Determine the potential system payback period and ROI.

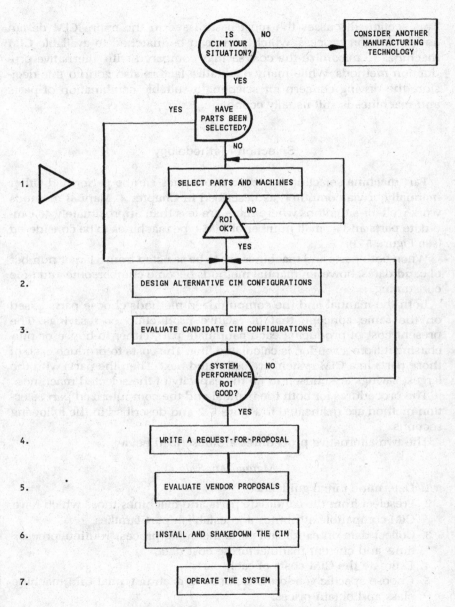

Figure 5-1.   Acquisition of CIM: STEP 1

- Preselect parts and machines having CIM-compatible attributes
- Preselect parts and machines having CIM-compatible attributes from available candidates.
- Calculate current production cost of each part.
- Estimate CIM costs for each part.
- Select the most economically beneficial parts and machines.
- Perform investment analysis to determine if a CIM system is an economic alternative.
- Data required:
  1. For each candidate mechanical part:
     - Machining cube size.
     - Material.
     - Cost to buy part from vendor (if a "buy" part).
     - Annual production volume.
     - Current machine types used in manufacturing sequence.
     - Set-up time on each machine type.
     - Run time on each machine type.
     - Manufacturing cost for each unit of setup and run time (direct labor and overhead).
     - Unique identification of each tool.
  2. Cost of specific machines in each machine class, approximate cost of an MHS, computer, tools, fixtures, etc.
  3. Machining cube of each CIM system elements by each machine type.
  4. Estimated CIM costs per part per unit of run time.
  5. Number of available production hours annually.
  6. Desired system(s) size(s)—governed by projected amount of capital available for CIM purchase.
  7. Total investment for each specified system size.

Figure 5-2. Step 1—Select Parts and Machines

8. Choose a different set of machines and repeat the investment analysis if necessary.
9. Repeat steps (6), (7), and (8) several times to obtain the most cost-effective solution which meets your requirements.

Computerized Analysis

1. Determine initial guidelines.

2. Preselect from the candidate parts and machines those which have CIM-compatible attributes.
3. Collect data on candidate parts including process routing, process time, and current manufacturing cost data.
4. Estimate the CIM costs of each part.
5. Choose specific vendors' machines in each potential CIM machine class and obtain prices.
6. Select machines and parts.
7. Determine the potential system payback period and ROI.

The iterative nature of the manual part machine selection method is an attempt to find the best solution among the candidate parts and machines, but usually the number of combinations is sufficiently large that all of them cannot be examined. Thus, the resulting manual solution is not necessarily the "best."

In contrast, a computerized part machine selection approach finds an optimum combination of parts and machines. The computer-based selection algorithm can estimate the relative savings of all the candidate parts. It also considers for each part the size of machine required, the amount of machining time used on that machine, and the amount of machining time remaining on that machine. Both parts and machines are selected simultaneously, and the approach assures that the best possible combination was chosen.

Initial Guidelines

This section discusses types of machines used in the CIM system as well as other decisions that must be made before a final determination can be used.

A basic issue is the maximum size of the system to be considered. This may be determined by floor space availability, budgetary constraints, or simply by the size of the system that the available personnel handled. A second issue is the total annual operating time of the CIM system. As with stand-alone NC machines, a CIM system is capital intensive and, therefore, the longer it operates per year, the better the ROI. This logic suggests a three-shift, seven-day-per-week operation. In addition, the level of production accuracy usually increases as the period between shutdowns is lengthened. Many factory environments will not support three-shift operation, however. People, in general, do not like to work the third shift, and there is little time for preventative maintenance. Often, five-day-per-week, two-shift operation is more appropriate.

Also at issue is what class of parts to produce on the CIM system:

prismatic, rotational, or a combination of parts. CIM technology for prismatic parts (flat or boxlike, basically rectangular, solid parts) is well established. The technology for rotational parts (shafts and disks), especially between-centers shaft work, is still relatively new. The turning of disk-shaped parts presents less of a problem, as these parts can be fixtured for vertical turret lathes. Each of these part classes will have its own design peculiarities and problems. The choice should be based on the available work content as well as how comfortable the organization is with the available technology. It will be necessary to choose between currently produced parts, completely new parts, spare parts, or some combination of these.

The range of part sizes to include depends on a number of factors. The most obvious is the average size of the parts. Large parts dictate large machines. Small machines are usually better for small parts except when several small parts can be mounted on a single fixture.

Small (6-inch machining cube or less) parts usually have a small amount of work content, and may only need to be at any given machine for a few minutes. The shorter the cycle time in a CIM system, the more parts that must be in the system at any one time. This may overload the material handling system (MHS) and degrade overall system efficiency. Large parts (36-inch machining cube or larger) usually require special handling consideration, since neither their size or weight (when combined with a fixture and pallet) are easily handled by currently popular MHSs. Also, as part machining cube size increases, so does the cost of the machines to be used. Thus, there is a tendency to limit machining cube size to less than 36 inches and greater than 6 inches. This allows the parts to be transported using conventional material handling systems and can permit multiple loading of small parts on one large fixture, to lengthen the total time at each machine and reduce the possibility of MHS bottlenecks.

Preselection of Parts and Machines

Out of the total set of parts of potential interest, it should be possible to preselect a subset suitable for manufacture on a CIM system. Similarly, out of the total of CIM machines available, it should be possible to identify a subset suitable for the set of candidate parts. Thus, the field is narrowed and the problem of final selection simplified.

If the candidate parts have been coded using a group technology classification system, then the fastest and simplest approach to preselection is to sort parts using the computer, based on CIM-compatible part attributes. Typical attributes include:

- Desired machining cube;
- Material (e.g., aluminum or steel);
- Form (prismatic solid box, disk, flat);
- Types of operations (milling, drilling, boring, etc.);
- Tolerances;
- Production quantity;
- Machining time;
- Current number of fixturings.

These few criteria can usually greatly reduce the candidate part set with little effort. One company has recently completed the installation of a group technology classification and coding package and has processed approximately 3,300 different parts. Sorting those parts for CIM-compatible attributes revealed 677 parts that were promising candidates.

Without a group technology system, sorting manually by reviewing part documentation and process plans can consume a great deal of time and personnel. A benefit of an automated part machine selection algorithm is that, although preselection is useful to eliminate unsuitable candidates and reduce the part set to be examined (saving computational time), a selection algorithm can review all of the candidate parts in a very short time and does not require group technology. However, the results must be reviewed in detail to make sure none of the unsuitable parts were chosen because of misleading information. For example, one manufacturer machined one end of a 10-foot long part on a small milling machine. Although the part was too large for a normal system, without sorting the parts based on size, all that was known about the part was that it was machined on a small milling machine and, therefore, acceptable for CIM production. If some selected parts are unsuitable, they can be removed from the candidate part set and the selection program rerun. This iterative process is continued until no unsuitable parts are chosen.

The number of CIM fixturings per part can also be used to preselect candidate parts. Although it can vary with system application, a rough rule of thumb is that if the part must be fixtured more than three or four times, it should be rejected. Again, this is done automatically by the computer program.

Another issue concerns tool wear. A part which requires extensive work on a hard material will wear out tools rapidly and impose excessive requirements for tool replacements. Such intervention interrupts the CIM system and interferes with productivity. Clearly, a part requiring a large number of short operations would be preferable to one requiring a few very long operations.

One method of preselecting parts, often used in conjunction with the manual part selection approach, is to group the parts into families. Grouping parts in this manner emphasizes the similarities of those parts, and one or two of the families can be chosen for manufacture on the CIM system. Three common methods of grouping parts into families are:

- Assembly: All the parts necessary to produce some end item or sub-assembly.
- Size and Common Manufacturing Operations: A number of parts that require approximately the same machining cube and parts which require the same types of machining operations—milling, drilling, tapping, etc.
- Type: All parts of the same type, e.g., transmission housings.

Often, some of the part families contain enough work content for a normal size CIM system, and the decision will be to decide which one would be best. The drawback to this approach is that the most economic combination may include parts from a number of families. However, if each family consists of many parts, it may be possible to choose the family which has the most potential for CIM savings, and continue the preselection and selection processes on that family only.

The preselection of machines depends more on preferences than hard constraints. The class(es) of parts chosen will limit the classes of machines, as will the range of machining cubes chosen. Average accuracy requirements must also be considered, as well as part materials. Finally, familiarity with certain types of machines, say horizontal machining centers rather than vertical ones, can also be used to limit the number of candidate machines.

The next result of the preliminary planning pass for the CIM system is a feasible set of parts and a set of machines from which everything unsuitable has been removed.

Data Collection

For both manual and automated part machine selection, the next step is to construct a data base (with the following components) for the parts/machines that are now under consideration:

1. Process-routing and operation data. At the planning stage, the following data will be required for each candidate part:
   - Routing Sequence—The machine classes (e.g., lathes, machining

centers—standard-precision, high-precision,—etc.) the part must visit to be machined, and the proper manufacturing sequence.
- Estimate of total process and fixturing times.
- Fixturing concepts and fixturing times.

2. Current Manufacturing Costs. Manufacturing costs can be estimated from the current cost of buying the part or from the components of in-house cost (direct labor, overhead, etc.) for each candidate part. Alternatively, the hourly machine rate cost used to quote jobs can be used. Neither of these cost concepts should include any reference to capital recovery costs or depreciation; they should strictly reflect daily operation cost. The manufacturing cost per part is then simply the machining time per part (MT) plus the set-up time (ST) prorated over the part batch size (BS) multiplied by the machine rate (MR) or shop labor rate (DL) multiplied by the applied overhead (OH), all multiplied by the part's annual production volume (P). That is:

$$\text{ANNUAL MANUFACTURING PART COST} = MT + \frac{ST}{BS} \times MR \times P$$

or

$$= MT + \frac{ST}{BS} \times DL \times OH \times P$$

CIM Costs

To be able to estimate the savings due to the use of CIM, the approximate annual CIM costs must be known to apply to the time an assembly will be in the CIM system. Nominally, this time is the cumulative machining times plus the load and unload time required for each fixturing of the part, multiplied by the part's yearly production requirement. This cost can only be approximate due to the assumptions made as to manpower requirements, machining times, and load/unload time (fixturing time). Based on system size, it is possible to roughly estimate the manpower requirements. Required are a system manager, 0.25 electrical technician, and 0.25 mechanical technician for every four machines in the system. Depending upon the number of tools that might be required in the system (60 tools per machine is a good rule of thumb), 0.5 to 1.0 tool setter will be needed for every five machines. Finally, if part cycle times are short or there are many fixturings for each part, at least two loaders will be required.

When estimating manpower requirements, the labor rate and overhead to apply to the system should be determined. In the simplest case, conventional direct labor cost plus overhead for the machine rate can be used. However, this usually does not provide sufficient overhead allocation because the overhead rate was based on the assumption of one man per machine, which is not the case with CIM. It is often necessary to work with the accounting department to develop the direct labor rate and applied overhead rate for the particular situation.

Assumptions about machining time are easier to make. In the simplest case, assume that the cycle times in the CIM system equal those in the conventional method. This is reasonable if parts are currently produced on NC machining centers using palletized fixtures and pallet shuttles. If standard NC machining centers and job shop type temporary fixtures are used, up to 25 percent of the cycle time can be expected to be saved by changing to CIM. In the case of conventional manual machines (especially if the equipment is old), a 50 percent savings in cycle time may be reasonable.

Fixturing time is the easiest of the assumptions to make. Although in reality the number of fixturings varies with each part, assume that, on average, each part requires two fixturings and each fixturing requires eight minutes—five minutes to load and three to unload.

For the computerized selection method, an estimate of the amortized cost of fixtures is needed. This must be done for each part, as fixture complexity can vary widely from part to part. The amortized cost is a yearly value which represents the annual cost to the company for buying that fixture, based on some estimated rate-of-return (i) at which the company could have invested the money in some other project. Fixture prices range from approximately \$4,000 for very simple fixtures to \$25,000 for very complex window frame fixtures and pedestal fixtures. The number of fixtures and their costs are estimated and then amortized over the production life of the part at a chosen corporate rate (i). As an equation:

AMORTIZED (ANNUALIZED)

$$\text{COST } (AC) = \frac{\text{TOTAL FIXTURE COST} \times i \times (1 + i)N}{(1 + i)N - 1}$$

where:

N is the part's production life.

This completes the basic elements of CIM costs. The annual manufacturing cost per part on a CIM system is then:

ANNUAL MFG PART COST $(AMPC) = (MT + FT) \times P \times W \times (DL \times OH) + AC$

where:

MT = Machining time
FT = Fixture time
P = Annual production rate
W = Number of workers
DL = Direct labor rate (in the same units as production time)
OH = Overhead rate
AC = Amortized fixture costs

CIM Machine Cost Data

It is now necessary to choose actual machines for each chosen machine class and determine their cost. This involves going to various vendors for quotes on the machines. It is best to choose a number of vendors in each class, as well as a number of sizes in each class.

To use the automated part and machine selection algorithm described as follows, it is necessary to know not only the cost of the machine but the amortized cost of that machine plus the remainder of the system (MHS, computer control, tools, etc.). This is accomplished by estimating the cost of the remainder of the system (based on system size) using the component costs from CIM vendors, amortizing that cost over the expected life of the system, amortizing the cost of the machine over the same period, and adding the two annual costs together. The amortization equation is the same as for fixture amortization, but the concept of salvage value now comes into play. At the end of the useful life of the equipment, it may be of value to someone else and it can be sold. The equation now becomes:

$$\text{AMORTIZED COST} = \frac{(IC - SV) \times i \times (1 + i)N}{(1 + i)N - 1} + (SV \times i)$$

where:

IC = Investment cost
SV = Salvage value at year N
N = Production life
i = Investment interest rate

For each chosen machine, the value used in the program equals the amortized cost of the machine plus the amortized cost of the rest of the system (a constant for the size of system chosen).

Part and Machine Selection

At this point, the current cost to manufacture or buy each part is known, as well as the estimated CIM costs for each part. Also, specific prices for machines in each CIM machine class (or an average price for a representative machine in each class) have been obtained. The next step is to actually select parts and machines for the CIM system. Although the selection theory is the same, its implementation differs for the manual and automatic selection methods. The procedure for manual selection will be discussed first, followed by the automatic selection method.

To begin the manual selection process, a set of machines must be chosen from the available machine classes. This selection can be arbitrary, or it can be based on the current production equipment used to produce some of the parts. The number of machines from each machine class is somewhat arbitrary; however, the total number of machines should equal the maximum set by management at the beginning of this process.

Choose parts to load these machines in the following manner. For each part, subtract the CIM costs from the conventional part cost to obtain a value for the cost savings for each part. Then begin loading the machines with the parts having the largest values. Base the available machine time on annual production hours available (two or three shifts, five days a week, fifty weeks per year) multiplied by an availability or efficiency factor (usually between 0.65 and 0.8, or 65 to 80 percent) to account for downtime, preventative maintenance, etc. The part cost savings and production time is already in annual values, so questions of sufficient aggregate production volumes are answered positively if the machines can be fully loaded. If two parts have equal savings, the one using the least machining time is usually better, so compare the machining times as well as the savings when choosing parts. Continue to choose parts until all machines are loaded or there are no more parts. Add up the savings for all of the parts chosen.

Configuration Design Issues

When the parts and machine types have been selected, it is possible to proceed to the system design process. There are several issues involved:

- Flexibility;
- MHS configurations;
- Machinability and process planning;
- Required accuracy;
- Required system availability;
- Tool-changer capacity;
- Other processes (inspection, heat treatment, finishing, etc.).

Figure 5-3 describes the acquisition of CIM (implementation step 2) and Figure 5-4 summarizes the steps.

Flexibility

Flexibility in a CIM system has many aspects. The most important of these is the random-processing capability which allows more than one part number in the system at one time. Usually, ratios of part types to one another can be arranged to meet current production needs, allowing rapid adaptability to changing market requirements. A CIM system is also relatively insensitive to engineering design and tooling changes.

Some CIM systems exhibit flexibility to fault tolerance. They continue to operate almost normally in the presence of machine failures, with other machines "covering" for the ones out of service.

A third type of CIM flexibility is the ability to operate virtually unattended. Maintenance and part fixturing can be performed during the first shift, while much of the actual production occurs during the second and third shifts.

Desired flexibility affects CIM design. Untended second- and third-shift operation implies some automatic, in-line inspection. Very high system availability implies not only reliable machines, MHS, and computer but possibly redundancies, such as back-up computers, duplicate machine types, and alternative routings in the MHS.

Flexibility affects other design features. If a CIM system is expected to handle a very wide range of part types, the machines will have fairly general characteristics such as multi-axis capability, reasonable precision, more than minimum horsepower, etc. A requirement to accommodate several part numbers simultaneously (without batching) implies large tool-storage capacities.

A related aspect of flexibility is the expandability of the system to accommodate future increases in demand or allow a "phased" installation—to maintain machine utilization over time and/or to prevent overtaxing the available capital. The degree of expandability desired

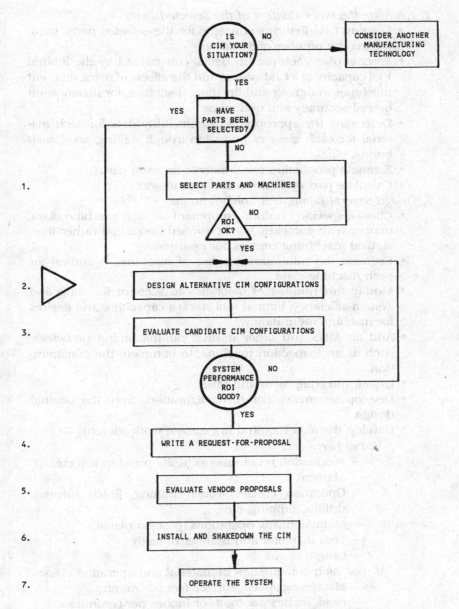

Figure 5-3. Acquisition of A CIM System: STEP 2

1. *Estimate the Work Content of the Selected Parts*
   - Develop CIM fixturing concepts for the selected parts, minimizing the number of fixturings.
   - Process plan each part in detail, constrained by the limited tool capacity of a CIM system and the effects of using different machines (roughing and finishing machines, for instance) on overall accuracy and cycle time.
   - Determine the appropriate machinability data for each material for each class of operation (rough milling, semifinish boring, etc.).
   - Estimate production requirements for each part.
   - Calculate part cycle times and tool usage.
2. *Design Several Equipment Configurations*
   - Choose specific vendors' equipment in each machine class; temper with company biases (toward horizontal rather than vertical machining centers, for example).
   - Estimate the minimum number of machines (spindles) for each machine class.
   - Modify this number of machines to account for shop and system efficiency, limited tool storage capacities, and desires for machine redundancy.
   - Add an MHS and other desired nonmachining processes, such as an inspection machine, to complete the configuration.
   - Layout the equipment and MHS.
   - Develop alternative design configurations from the original design.
   - Develop the data required to estimate work content:
     a. Per part:
        — Production requirements (units per ship set, etc.).
        — Material.
        — Operation classes (rough milling, finish turning, drilling, tapping, etc.).
        — Manufacturing operations (process plans).
        — Tool diameter and number of teeth.
        — Length of cut.
     b. For each combination of material and operation class:
        — Machining speed, surface feet per minute.
        — Feed, inches per tooth or inches per revolution.
        — Machinability rating or power factor.

Figure 5-4. Step 2: Design Alternative CIM System Configurations

c. For each of the potential CIM machines:
  — Maximum spindle speed, revolutions per minute.
  — Maximum tool feed, inches per minute.
  — Horsepower.
  — Chip-to-chip tool-change time, seconds.
  — Average rapid traverse time between repeated operations, seconds.
  — Transfer time.
* Rotary table speed, degrees per second.
* Pallet shuttle time, one way (on or off machine), seconds.
* Data required for configuration design:
  a. Specific equipment from different vendors in each machine class.
  b. Options on that equipment, such as maximum toolchanger storage capacity, pallet shuttles, spindle probes, broken tool sensors, adaptive controls and diagnostics.
  c. Accuracy of each machine.
  d. Specific MHSs. (Most often, each machine vendor will have one or two MHSs available, matched to his equipment.) Specific inspection equipment, wash stations, tool-room equipment.
  e. Shop efficiency factor

Figure 5-4. *(Step 2 continued)*

will affect the choice of MHS and the arrangement (layout) of equipment in the configuration.

Flexibility is desirable, but it may increase cost. By the same token, too many special-purpose machines will hamper a system's ability to handle new part types. Yet, specialized machines, such as multiple-spindle head-changing units, may be needed to increase system throughput, thereby reducing the cost per part. Since future requirements can never be predicted with certainty, it is necessary to rely upon judgment, not just a cost/benefit analysis, to resolve the issue of flexibility.

Machinability and Process Planning for a CIM System

Basic data about how the workpiece can be processed is crucial. When treated lightly, it has caused serious problems in several installations. One company overestimated feeds and speeds, resulting in poor surface finish, rapid turnover of tools, and finally, a 25 percent reduction in system throughput. Another company used the wrong material condition when obtaining machinability estimates from tables in an out-of-date handbook. Compounded by some other data mistakes, this error underestimated the cycle times of the parts by one half to one third, and correcting the situation resulted in a reduction in system throughout of 75 percent.

A logical approach to defining machinability data is to first obtain the optimum feeds and speeds for the part materials for each tool category from a current machining data handbook. Then, the machine operators should be asked what feeds and speeds they routinely use when machining those same materials. The CIM system feeds and speeds that should be used will fall between these two values, adjusted for part rigidity, the ability of dedicated fixtures to hold the parts better than standard component-built fixtures, and the use of new machine tools, perhaps with adaptive control to maintain optimum cutting parameters.

During the CIM configuration design, a change in feeds or speeds for tools necessitates recalculating machining time for all the parts using that tool; if there are many parts, this revision process can become extremely time-consuming (if done manually), especially if many alternatives are evaluated. This can be simplified by using a computer program to calculate cycle times. Where the metal's characteristics are well known, it is better to assess the accuracy of machinability data before beginning configuration design, temper it with normal shop practice, and attempt to hold that data constant throughout the configuration design exercise.

Process planning for CIM is different than process planning for standalone machines. There are two areas of critical importance: the fixturing approach used for each part and the selection of cutting tools. In a CIM system, it is important to minimize handling of the part. Careful attention to fixture designs can help. The use of window-frame and pedestal-type fixtures allows access to a part when using either four- or five-axis machining centers. (Try to use four-axis, X, Y, Z, and rotary-table machining centers whenever possible. The cost of five-axis machining centers can seldom be justified by the parts' work content that might require a tilting table or spindle.)

The goal is to have one fixture type per part type, in order to have only one load/unload sequence. While this is seldom obtainable, a considerable amount of time should be spent analyzing each part to determine its best fixturing attitude with respect to the required operations and the machines. Then the fixture should be designed to maintain that attitude. Also, by matching machine axes, it is possible to minimize the number of fixtures and fixturing. For example, if vertical turret lathes (VTLs) are being employed for the rotational work content, investigate the use of vertical machining centers for the prismatic work content.

Tool-storage capacity needs can be minimized by standardizing the tool selection during process planning. This is most noticeable in the choice of milling cutters. For example, a standard 2- or 3-inch diameter shell mill could be used for all milling operations, except where corner radius or pocket size may not allow it. Also, compound boring bars should be investigated, where more than one diameter and/or chamfers may be cut at the same time. These are "specials" and priced accordingly, but they save time in tool changes and the cost of the additional single-tip bars. Plus, they may provide better accuracy. Though more difficult, it may be possible to standardize drill and tap sizes and hole patterns throughout the part set, again reducing the number of tools to set, store, control, and maintain.

Changes in process plans can require additional tools, refixturing or rotating of the part, or may even require a special machine in the system. As with machinability data, make sure process plans remain as constant as possible throughout the design phase.

Precision

With the recent trend toward tight tolerances (for example, 0.001 to 0.0001 inch), otherwise typical parts can require accuracy unavailable from general-purpose machines normally well-suited for inclusion in a CIM system. The CIM buyer/designer is faced with a number of alternatives.

The first alternative is to request an increase in tolerance, although it is unlikely that all of the offending tolerances will be relaxed.

If high-precision machinings are required, consideration should be given to the advantages and disadvantages of producing those features off-line. When evaluating the off-line option, consider the additional time required due to transportation, the control problems resulting from a part leaving the system and possibly returning, whether the part will have to be removed from its fixture/pallet, and the cost of purchasing a high-precision machine, which may be underutilized.

If it is decided that the work should be done on-line, high-precision machines must be included in the system. Compatibility problems can be minimized if the system vendor also supplies the special machine. Alternatively, the supplier may be willing to customize general-purpose equipment to obtain the desired accuracy, if the cost can be justified. However, problems might arise from this approach; the untried design may have to be debugged as it is brought on-line, delaying production.

Finally, if the vendor will not build high-precision machines, the prime vendor may obtain one elsewhere and integrate it into the system.

If on-line machines are to perform only high precision machining, they may well be underutilized and add to the cost of the parts. If, however, normal machining work is assigned to the high-precision machines, care must be taken to prevent its overloading, possibly affecting the machine's basic accuracy.

Finally, consider the need for environmental control to maintain a stable atmosphere with which to obtain the desired accuracy. This may include temperature control of the CIM area, part/fixture/pallet temperature soaking, and coolant temperature conditioning. It may also mean temperature control of the inspection equipment.

Uptime

Machine availability (usually expressed as a percent) is the time during which a machine is not failed, that is, the time it would be processing a part if a part were available. This availability is also called the "uptime" percentage. The uptime of the entire CIM system would be related to machines not failed or machines by-passed successfully keeping the system operational.

The time to repair a machine or an entire CIM system is composed of several factors:

1. Time required to discover the failure.
2. Time required to call a maintenance person.
3. Time required to diagnose the failure.
4. Time required to obtain replacement parts.
5. Time required to actually make the repair and test for proper operation.

Consequently, there can be considerable variation in the repair times. The average or mean time to repair (MTTR) a machine in an established CIM system is likely to be in the range of 0.5 to 2.0 hours. The MTTR and the mean time between failures (MTBF) for a machine defines the machine's average availability figure.

System availability is the percentage of time that none of the system components are failed, that is, all machines, controllers, computers, the MHS, etc., are "up." Clearly, the average availabilities of the components of a CIM system must be high if the average system availability is to be high. For example, if each of 10 components of a CIM system has an average availability figure of 98 percent (and if the component failures are assumed to be always independent of one another), then the average availability of the system will be approximately 82 percent (that is, 0.98 × 0.98 ten times). Usually, however, one or two of the components have availability figures lower than the rest, and these figures are then the primary determinants of the system availability.

Of course, the failure of one machine does not always mean that all production ceases until it is repaired. In a well-designed flexible system, production can often continue (at a reduced rate) while portions are under repair. Parts that normally visit the failed machine might be rerouted to other, similar ones if the appropriate tooling and part programs can be made available. Or, the affected operations might temporarily be performed off-line, for example, manual inspection in the case of a failed automatic inspection station. In the case of failure of the central computer controller, the machines could be operated semiautomatically from their own controllers.

Therefore, it is important to not only specify highly reliable system components, but to also anticipate failures. A common design method is to incorporate redundancy in the system. This may mean that the system contains two or more machines of each type. Or it may imply machines backing up dissimilar machines, for example, a machining center might substitute for a multiple-spindle machine. Redundancy in the MHS implies multiple part carriers (for example, carts) and multiple paths between stations (in case certain links should fail). Obviously, there are tradeoffs between system redundancy, system complexity, component reliability, capital cost, and the cost of lost production.

Tool Storage Capability

Machines with large tool changer storage capacity are generally chosen because they reduce the need for production batching and they can facilitate the processing of parts rerouted from failed machines.

Other Processes

How many processes, in addition to machining, should be done on-line and how much should be done off-line? These other processes include very high accuracy machining, washing, inspection, stress relieving, heat treating, deburring, finishing, marking, and assembly. Except for washing and inspection, in general these processes should be kept off-line. A rule of thumb is that if the part must be removed from the pallet/fixture before starting an operation, that operation should be done off-line.

However, considering the control problems created by sending parts off-line (especially if the parts must be removed from their fixtures) and then returning them, in certain cases the cost of providing on-line equipment for some of these processes is justifiable, though the processes need not be automated. For example, a manual inspection station could be on-line.

Material Handling Systems (MHS)

The simplest MHS consists of a person and a cart, manually moving palleted parts from machine to machine under computer direction. To reduce machine waiting times, a shuttle loader can be added to each machine tool. This manual system will work for small CIM systems, where the distance between machines is short and where parts are relatively small and light. However, for larger systems and/or heavier parts, automatic MHSs are more applicable. These systems consist primarily of carts, conveyors, or robots that carry pallets automatically to and from each shuttle loader. If the loader is full, the pallet will circulate in the MHS, wait in front of the machine, or go to an off-line buffer storage area. At present, a person is usually required to fixture and defixture parts at load/unload stations (the Japanese have successfully used robots in some applications), but the rest of the system is under direct computer control.

Configuration Design Procedure

The design of a CIM system configuration is a four-step process. The steps are:

1. Based on the work content of the parts or the recommendations of the part/machine-selection algorithm (if used) and company biases (toward horizontal-spindle rather than vertical-spindle machining centers, for example), choose specific vendors' equipment in each required machine class (horizontal or vertical lathes or machining centers, standard or high-precision machines, etc.) with sufficient horsepower and accuracy to produce the parts selected.
2. Using the work content of the selected parts, production requirements, available work hours, and data about the chosen equipment, estimate the *minimum* number of spindles—or the number of machines in each class—necessary to obtain the production goal.
3. Modify (increase) the number of machines in each class, including inspection machines, by taking into account machine and sys-

tem efficiency, limited tool capacities, desires for redundancy, and machine loading decisions with respect to any high-precision machines.
4. Add an MHS, wash stations, and other nonmachining processes to complete the configuration design.

From this point on, the design process is more aptly named the "design and evaluation process," because of its iterative nature, that is, design, evaluate, redesign, etc.

Configuration Design Example: General Electric CIM System

General Electric purchased eleven machined parts from outside sources. The following paragraphs summarize the results of a study performed to determine the advantages and disadvantages of producing these parts in-house on a CIM system.

The parts are aluminum and have numerous thin cross sections. They are prismatic, can be machined in a 24-inch cube, and have an initial production rate of 600 each per year. These attributes led the GE staff to consider purchasing a CIM system to produce the parts.

The first step in designing a CIM system was to develop the work content (process plans) for each part. Planners at GE found that developing the process plans for all 11 parts and recalculating them for changes in machinability or machines would be tedious. A machine cycle time calculation program (CTIME) was created incorporating standard feed, speed, and time equations, yet allowing changes in machinability data, machine parameters, and machining elements. It facilitated rapid examination of the effects of such changes, especially during machinability experiments and when comparing different machines with different automatic tool changing times.

To minimize set-up time and the number of setups, the planners assumed the parts would be machined in "window frame" fixtures mounted on pallets, allowing access to at least four sides of the parts. The planners then entered the machining elements for each part into the computer, along with tool information, machinability data, and average machine tool data. The CTIME program estimated the cutting time, dead time (tool change, travel, part rotation, etc.), and total time for production, as well as the machine spindle speeds and feeds.

They found that, assuming the equipment would be used two shifts during each workday, there were two spindles' worth of work content at the present production rate for the 11 parts. Additionally, approximately 30 percent of the work content required high-precision machining (tol-

	Machine	O GP Mach Center (1) Precision Boring Mills (6)	A GP Mach Center (1) Precision Boring Mills (4)	B GP Mach Center (1) Precision Mach Center (1)	C Precision Mach Center (2)
	Material Handling				
MANUAL	I Manual	Job Shop			
	II Carousel Loader	(N.A.)			
AUTOMATED	III Line (conveyor or cart)	(N.A.)			
	IV Loop (conveyor or cart)	(N.A.)			

Bolted Fixtures · · · · · · Palleted Fixtures

Figure 5-5.   CIM Configuration Alternatives

erances of less than ±0.0005 inch). This information was used to develop the configuration alternative matrix illustrated in Figure 5–5. The types of MHS are shown on the vertical axis and the types of machines are shown on the horizontal axis. Thus, each box represents a CIM configuration consisting of specific machines and a specific MHS.

In the matrix, configuration I,O is a common job shop and is the present manufacturing method. The other alternatives represent compromises in CIM design issues, such as precision, redundancy, and manpower reduction. Column A consists of configurations with one NC general-purpose machining center and a number of manually operating precision boring mills. Note that the required number of boring mills is less in column A than in column O because time is saved by setting up the parts on pallets rather than on the machine beds.

Some of the obvious characteristics of this alternative are as follows: System maintenance requirements are principally for standard mechanical failures of the machines; little electronic or NC skill is required. At least four precision boring mills are needed to satisfy production requirements, resulting in high capital expenditure to perform 30 percent of the work content. Significant time will be used transporting parts from the general-purpose NC machine to the high-precision machines.

Finally, while there is redundancy in the manual boring mills, none exists for the NC machine. If that machine fails, production will cease or the high-precision machines will be required to do general work, greatly reducing production and possibly damaging the machines.

Column B consists of configurations with one NC general-purpose machining center and one NC high-precision machining center. These configurations share many of the problems of those in column A. Manpower is reduced (fully automatic operation is feasible), but there is no redundancy, and high accuracy is more difficult to maintain without manual intervention. Redundancy can be created by adding a second machine of each class, but that would be extremely expensive since two machines can handle all the work. Compatibility of equipment becomes a concern, since not many manufacturers produce both standard and high-precision machining centers. Controls, pallets, and tooling can be different, and often one manufacturer will hesitate to interface another's equipment with his own. Maintenance requirements also increase.

Column C consists of configurations with two high-precision machining centers, and the configurations assume that the machines are sturdy enough to be unaffected by roughing work. If this assumption holds true, many of the design problems mentioned previously are solved. There is full redundancy, no compatibility problems, little manpower required, and no transportation time from machine to machine. Accuracy is still more difficult to obtain than with manual intervention, however, and maintenance still requires highly skilled electronics and mechanical workers.

To each of these configurations, various methods of material handling would be added. Inspection equipment, wash stations, and so on, complete the designs.

Evaluating Candidate CIM Designs

Having previously discussed configuration design for a variety of CIM configurations, this section describes how these configurations can be compared with respect to cost, throughput, reliability, accuracy, and ROI. This is Step 3 of the CIM implementation process (Figure 5–6, Figure 5–7.)

Evaluation of a number of alternative CIM configurations should be performed using a systematic, step-by-step approach. A systematic evaluation technique is presented in this section; the steps may be summarized as follows:

1. Construct an evaluation matrix showing all factors considered important to the evaluation of each configuration.

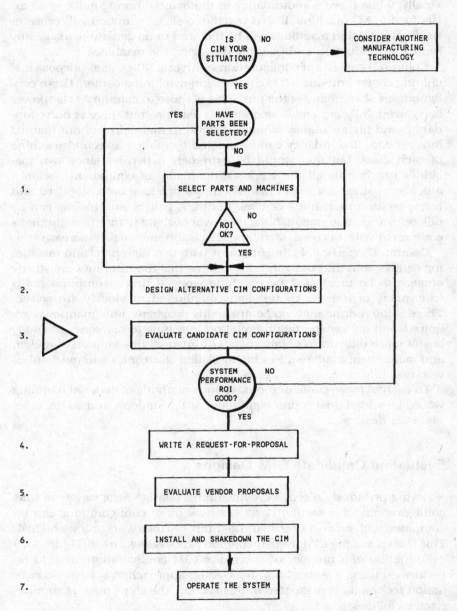

Figure 5-6. Acquisition of A CIM System: STEP 3

- Simulate the operation of each configuration based on predetermined scheduling, batching, and balancing rules to provide performance measures for each configuration.
- Improve the configuration designs until each provides satisfactory performance measures or is rejected.
- Perform a detailed investment analysis of each configuration.
- Examine and evaluate intangibles, such as flexibility, accuracy, etc.
- Choose the configuration which best satisfies the investment and intangible analyses.
- Data required for configuration evaluation:
  1. Per part:
     — Routing between machines (primary and alternatives).
     — Run time at each machine.
     — Number of unique fixture types.
     — Quantity of each fixture type.
     — Number of tools at each machine.
     — Inspection time.
     — Fixturing and defixturing times.
  2. For simulation:
     — Part data.
     — Layout of configuration—number of stations for each machine type, load/unload, inspection, wash, and storage.
     — Part routings (primary and alternative) based on batching, balancing, scheduling strategies.
     — Travel time between stations.
     — Production requirements.
     — Available production time.
     — System efficiency factor.
     — Material handling operations logic.
     — Failure rates per machine and MHS (optional).
     — Maximum number of pallets/fixtures in the system at one time.
  3. For economic analysis:
     — Part cycle times (from simulation).
     — CIM costs per unit time (direct labor, overhead).
     — Number of batches needed to meet production requirements.

Figure 5-7. Step 3: Evaluate Candidate CIM Configurations

— Number of workers used to change system over at end
of batch.
— Production time lost due to batch changeovers.
— Number of full-time workers in the CIM system.
— Total investment including:
— Machine tool costs.
— Inspection machine costs.
— Load/unload station costs.
— Material handling system costs.
— Computer costs.
— Fixture costs.
— Pallet costs.
— Cutting tool costs.
— Part programming costs.
— Engineering costs.
— Installation costs.
— Spare part costs.
— Depreciation schedule.
— Investment tax credit.
— Tax rate.
— Minimum acceptable ROI.
4. For intangibles:
— Weighting scheme: A subjective scale applied to the
evaluation criteria indicating their relative importance.
— Rating scheme: A scale (from one to five, for example)
showing the degree of compliance of any configuration
to any criterion.

Figure 5-7. *(Step 3 continued)*

2. Develop operational strategies—batching, balancing, scheduling,
and dispatching rules to provide realistic input data to the simulation and economic analysis.
3. Simulate the operation of the configuration, noting performance
measures for the economic analysis and for comparison to alternative configurations.
4. Improve the design or create a new design, repeating the steps in
the configuration design procedure.
5. Repeat steps (1), (2), and (3) until a variety of design configurations
have been examined.

```
                Poor     ---->      -Excellent
        Ranking:   1    2    3    4    5


               Not Important  ---->  Very Important

        Weighting:  1    2    3    4    5
```

Criteria	Weighting	Configuration A	B	C
Shipset Production Time				
System Availability				
Redundancy				
Flexibility				
Tool Capacity				
Cost/ROI				
Inspection				
Surge Capacity				

Figure 5-8. Evaluation Matrix: Configuration Design Evaluation

6. Examine the configurations with respect to investment analysis; estimate their ROI and payback period.
7. Complete the evaluation matrix, and choose the configuration having the best rating.

Evaluation Matrix

A very useful means of organizing the configuration is to construct a matrix such as that illustrated in Figure 5-8. Across the top of the matrix, list the configuration designs. Down the left-hand side, list the important evaluation criteria. For each criterion, determine a value that indicates that criterion's relative importance to the other criteria, say a number from one to five. This then becomes the criterion's weighting value. Considerable judgment must be used in selecting and weighting the evaluation criteria. It is important to achieve an organizationwide consensus as to the criteria and their relative importance. This involvement will also help gain organizationwide CIM system acceptance.

At the same time, a rating scheme should be developed to indicate the degree of compliance any configuration has to any criterion. This can also be a number from one to five, for instance.

Eight important criteria are:

1. Cost.
2. System throughput.
3. Predicted system availability.
4. System flexibility, particularly the ability to adapt to changing part types.
5. Effectiveness in handling problems of precision, if any.
6. Tool-storage limitations and the resulting requirements for batching.
7. Redundancy.
8. Surge capacity.

Each configuration will have two numbers associated with each criterion; the rating value and the "score." The score equals the weighting value of the criterion multiplied by the rating value of the configuration for that criterion. High scores are associated with high compliance with important criteria.

After the matrix is complete (all configurations have been evaluated), the scores for each configuration are summed. The configuration with the highest total score should be the most acceptable configuration.

The following sections attempt to illuminate the steps in the evaluation process by suggesting approaches to defining operational strategies for the configurations, simulating the configurations, analyzing the economics of each configuration, and finally choosing one of the alternatives.

Operational Strategies

Two concepts are important to the optimal loading of a CIM system: batching and balancing and scheduling and dispatching. These strategies are formulated in general at this point in the evaluation process to provide realistic machine loading data for simulation of each configuration.

Production Batching and Machine Balancing

Production batching, the division of production into subgroups or lots, is necessary when tool storage capacity limitations do not allow all the desired parts to be machined in a CIM system at one time.

Occasionally, balancing the workload on the machines may be so difficult that batching is required.

Balancing the workload on each machine tool attempts to maximize machine tool utilization as well as relieve or avoid potential bottlenecks in the system, with the intent of maximizing system throughput. Often, however, it will not be possible to balance everything. This is especially true in systems with different types of machines, for example, general-purpose machines and specials.

Balancing can also be difficult where a large number of tools are needed for certain parts. The division of work content and tool changer storage limitations are crucial. Additionally, if parts are required to visit a number of machines, the effects of transport time and MHS congestion may reduce system throughput.

Optimization software use greatly reduces the need for trial-and-error batching and balancing for simulation. At the completion of the batching and balancing exercises, specific parts and tools will have been allocated to specific machines in an attempt to maximize system throughput. This "optimum allocation" process is in reality iterative; the allocation provides realistic information for the simulation. The simulation results indicate the "goodness" of the allocation and may suggest modifications which can be resimulated.

Scheduling and Dispatching

The strategies for batching and balancing must be implemented systematically through scheduling algorithms and dispatching rules. These allow parts to enter and leave the system in the proper sequence at the proper time. Although the rules are discussed in greater depth later (for the actual operation of a CIM system), they must be determined at this point so that the simulations can be made realistic.

Simulation

After developing the strategies—both for batching and for scheduling—the second step in analyzing alternative CIM configurations is to obtain system performance measures. For general measures, for example, system throughput, average time a workpiece is in the system, station utilization, etc., queueing models are usually a fast, inexpensive choice.

When compared with detailed simulations, queueing models are typically accurate to within 10 percent. For greater detail and random failure analysis, simulation models must be used. Both models are driven by the operational strategies plus part data, machinability data, and system data.

Typical output from a simulation includes:

• Performance measures, such as system throughput or average time to produce one set of parts, machine and MHS utilization, etc.
• System bottlenecks.
• System's reaction to machine or MHS failures.

Choosing a Queueing Model or Simulation Package

Options include purchasing so-called "generalized" packages, using packages currently in the public domain, or developing your own. CIM vendor packages, if available, will be tailored to their individual systems approach, and the buyer may need a different one for each company that proposes a system.

Simulation software may have been created around a certain class of CIM systems, usually with a specific type of MHS (conveyor, tow line carts, or carts on rails). When used to simulate another class of systems, the results must be suspect. General CIM system simulation packages should be adaptable, ideally to evaluate a variety of MHSs.

System throughput, average production time per part set and the utilization of the machines, MHS, and workers are usually all generated by the packages. However, queueing models and some simulators cannot model the effect of failures on these performances measures. Also, they may not automatically indicate the number of functional units (machines, carts, fixtures) necessary for optimum throughput.

On the other hand, too detailed a model can be both expensive and complicated to use. The most desirable simulation or queueing model would be one in which the level of detail could be tailored to a particular need.

Simulation Example:
General Electric CIM Configuration Evaluation

This section summarizes the modeling and simulation of alternative configurations for the GE design example.

Three different simulation models were constructed in order to faithfully model the various MHSs. In all, 20 different configuration designs were simulated. The models were written in the discrete-event simulation language, Extended Control and Simulation Language (ECSL). In the design stage, part and machine data were first "preprocessed" using the CTIME program in conjunction with machinability data to calculate total operation times. A summary was used to batch the production and to balance each machine.

This information was then used with information on a particular MHS system to create an appropriate model for each of the alternative configurations. Simulations were performed for sets of 50, 60, and 90 parts for all 20 alternative configurations. A sample of the simulation output for one such configuration is illustrated in Figure 5–9. Output information was collected for each of the configurations and organized to facilitate the economic analysis.

Economic Analysis

In this section, the detailed analysis is performed to justify the system to upper management. The basic economic theory is reiterated and then illustrated through the use of a case study. Of course, different organizations use different accounting methods so the information here must serve as a guide only.

There are three basic categories of CIM system economic analysis: replacement, capacity expansion, and displacement.

In each of these categories, the annualized acquisition cost of a CIM system is compared to other manufacturing alternatives, based on the difference in part manufacturing costs and capital invested.

Replacement analysis, often referred to as "cost reduction analysis," examines the *replacement* of current machines and technology with CIM machines and technology. This approach is used primarily when intro-

```
                    2 MACHINES IN 1 GROUP

        ... IMPUT DATA ...
        SIDE    PART    GRP 1
          1       1      1642
          2       1      1149
          3       4      1668
          4       4      1416
          5       9       610
          6      11       394
```

THIS OUTPUT ASSUMES THE FOLLOWING: THERE ARE 240 PRODUCTION
HOURS AVAILABLE PER MONTH (300 ACTUAL HOURS WITH A MACHINING
EFFICIENCY OF 80 PERCENT). THREE DIFFERENT LEVELS OF
PRODUCTION ARE TO BE CONSIDERED: 50, 60, AND 90 SHIPSETS PER
MONTH. EACH SHIPSET IS MACHINED IN 3.0 BATCHES DUE TO THE
TOOL CHANGING CAPACITY OF THIS MACHINE CONFIGURATION.

38773 UNITS OF TIME FOR 10 SHIP-SET: ONE BATCH EACH TIME
PER SHIP-SET: 7753.9 (193.8 MIN).

```
TOTAL PRODUCTION TIME FOR 50.  SHIPSETS:  9693.2   MIN
PORTION OF AVAILABLE HOURS USED                .673  HRS
PRODUCTION HOURS LEFT:                        78.4   HRS

TOTAL PRODUCTION TIME FOR 60.  SHIPSETS: 11631.0   MIN
PORTION OF AVAILABLE HOURS USED                .807  HRS
PRODUCTION HOURS LEFT:                        46.1   HRS

TOTAL PRODUCTION TIME FOR 50.  SHIPSETS: 17447.0   MIN
PORTION OF AVAILABLE HOURS USED               1.211  HRS
PRODUCTION HOURS LEFT:                       -50.7   HRS

PROCESS WAS STARTED          59 TIMES
MOVE WAS STARTED            120 TIMES
BUFFER WAS USED               0 TIMES
UTILIZATION OF WORKER       .2228
UTILIZATION OF MOVER        .7427

... PRODUCTION OF PARTS ...
        SIDE    SCHED   COMP
          1       10      10
          2       10      10
          3       10      10
          4       10      10
          5       10      10
          6       10      10
        TOTAL     60      60
```

```
... PALLET UTILIZATION ...        ... STATION PERFORMANCE ...
      NO.    UTLZN                   NO.    GROUP   UTLZN
       1     .981                     1       1     .766
       2     .603                     2       1    1.000
       3     .927
       4     .550
       5     .593
       6     .760
```

```
... OPERATIONS COMPLETED ...
        SIDE    GRP 1
          1       10
          2       10
          3        9
          4       10
          5       10
          6       10
        TOTAL     59
```

Figure 5-9. Simulation Output

duction of a CIM system promises a significant reduction in manufacturing cost over the current method.

Capacity expansion (sometimes referred to as "cost avoidance analysis") examines the procurement of a CIM system instead of additional stand-alone machines to either manufacture a new family of parts or produce a greater volume of current parts. This approach is also used instead of replacement analysis when the current machines made available by introducing the CIM system can be used on other parts immediately.

Displacement analysis examines the displacement of current machines by a CIM system to provide additional manufacturing capacity sometime in the future. This approach can be used when no additional capacity is needed presently but will be needed in the future, shifting the analysis emphasis from replacement to time-phased expansion or cost avoidance. Experience has shown that CIM justification using standard cost-accounting procedures is most easily accomplished when additional capacity is required.

Two economic modeling techniques, each with its own advantages and disadvantages, can be used to perform the economic analysis regardless of the category chosen. The "Net Present Value" (NPV) technique estimates the present values of all of the savings and expenditures for the CIM system over its useful life, discounted back to the "present" by some value which represents the opportunity cost to a company for making that investment. In other words, if the company could presently invest at an interest rate of 18 percent annually, then the discount rate would be 18 percent. The "net present value" equals the present value minus the initial investment for the CIM system. If the net present value of the CIM system is greater than zero, then the CIM system is paying more than the discount rate on the money invested and should be implemented. If the value is less than zero, the CIM system is not economically justifiable.

The internal rate of return (IRR) technique estimates the discount rate at which all of the savings and expenditures for the CIM system over its useful life just equal the initial investment. It is exactly the same as the NPV technique, except that it estimates the discount rate instead of starting with it as a given. The project's IRR is then compared to a minimum acceptable value; if it is larger than that minimum value, the project is acceptable.

Both modeling techniques include the effects of taxes, depreciation, labor rate and material cost escalation, and any other cash inflows or outflows. Two major differences exist, however. First, the NPV technique makes an explicit assumption about the discount rate while the IRR technique has an implicit assumption. The discount rate is the assumed investment rate or reinvestment rate (like interest) at which the company could invest the income from the CIM system. The rate specified in the NPV technique is the investment rate that the company feels is likely for the service life of the CIM system. On the other hand, the IRR technique assumes that whatever discount rate equates the investment to the present value of the savings at time zero is a realistic reinvestment rate. However, as this discount rate begins to become much larger than the threshold rate, it becomes less likely that in reality the rate will be a feasible reinvestment rate.

If the discount rate generated by the IRR technique is of questionable merit, then why is this technique the more widely used? The answer lies in the second difference. The discount rate estimated by the IRR technique can be used as an index to compare different projects; it can indicate the best project on a relative scale. It is difficult to determine from the NPV technique which project makes the best use of the capital invested in it.

The basic methodology for economic analysis is to look at the increment of capital invested and compare it to the savings or costs resulting from it. For all categories of analysis, the manufacturing costs (direct labor, material, and overhead) to produce the parts must be calculated for both the conventional method, usually stand-alone machines, and for the CIM system. In cases where the parts were produced principally on manual machines, the first alternative should be to produce them on NC machines and then to compare both alternatives with the CIM system.

When performing replacement analysis, the increment of capital invested will be the investment for the CIM system (the salvage value of all equipment that the CIM system can eliminate is considered a cash inflow in Year One). This is then compared to the savings generated by using a CIM system instead of the machines replaced over the life of the CIM system or the project, whichever is less.

If it is an expansion analysis, the investment increment is the difference in the cost of purchasing the standard equipment necessary and purchasing a CIM system. This is again compared to the savings or costs of producing the parts on a CIM system instead of the conventional machines. Because of the higher utilization of machines in a CIM system, it is not unusual to find that the investment in a CIM system is less than that necessary to purchase the additional conventional equipment. In this situation, the CIM system can be beneficial even if it costs more to produce the parts on it.

Finally, for displacement analysis, the CIM system investment is compared to the investment over time of the machines that would have been required. This comparison, as before, is based on the savings or costs of producing the parts on a CIM system instead of the conventional machines plus a cost for storing and not using the conventional machine while it waits to be used. This opportunity cost of stored equipment is modeled as an additional cost due to the purchase of a CIM system and is subtracted from the annual savings of the CIM system.

If more than two alternatives are being compared, the economic modeling software automatically performs an incremental analysis. The alternatives are reviewed in ascending magnitude of investment and each increment of additional capital is either justified or rejected.

Economic Analysis Example: General Electric CIM Configuration Evaluation

Continuing the General Electric design example, this section summarizes the economic analysis of the configuration alternatives.

The investment analysis of the CIM system alternatives was done in two steps using an Investment Analysis Program (IAP). The first stage used a computer procedure that estimated the manufacturing cost (direct labor, material, and overhead) to produce a ship-set of parts using each alternative configuration. The second stage used another procedure that calculated the individual incremental-return-on-investments (IROIs) and payback periods both before and after taxes.

The first procedure, "Manufacturing Part Cost," estimated the manufacturing cost per set of parts for each configuration based on the number of full-time workers needed, the cost of a set's worth of castings, and a machine rate (a single dollar figure for direct labor and overhead). If no machine rate had been available, the program could have used the average hourly direct labor rate and applied overhead to it. Any other annual operating costs not included in the machine rate or overhead, such as special inspection gauges, were amortized over the year's production and added to the cost of each part set in order to estimate the manufacturing cost per ship-set. These input data are illustrated in Figure 5–10.

The second set of procedures, collectively called "Investment Analysis," calculated the IROI and payback period for each project, before and after taxes, using the NPR technique. The total installed cost was compared to the present worth of the net annual savings attributable to that configuration. (The net annual savings equals the cost of purchasing a ship-set from a vendor minus the manufacturing cost for the set produced on that CIM configuration. Costs for both vendor and in-house manufacturing were allowed to escalate annually at a predetermined

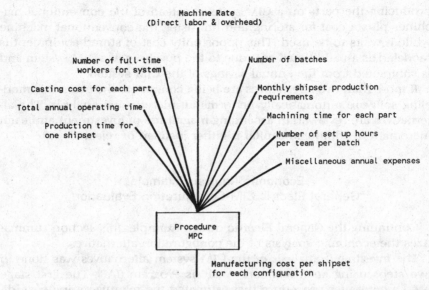

Figure 5-10.   Input Data to the Manufacturing Part Cost Procurement, MPC

rate, and the payback period was adjusted for the anticipated annual rate of inflation during that period.) The inputs and outputs of the investment analysis procedures IRR and NPV are shown in Figure 5-11.

In the case study, results of the economic analysis indicated that the fully automated three-machine (including an inspection machine) system was the most beneficial according to the criterion of choosing the largest investment with an acceptable IROI. General Electric was also interested in determining whether each increment of capital invested in the selected project above a "bare-bones" system of two DNC machining centers and manual material handling was justified. The analysis indicated that the fully automated system was justifiable.

Evaluation Matrix: General Electric CIM Configuration Evaluation

The evaluation matrix presented in Figure 5-12 provides an example of how the concept was applied to the General Electric configuration case study. Listed across the top of the form is a sample of the twenty configurations evaluated. Down the left-hand column are the evaluation

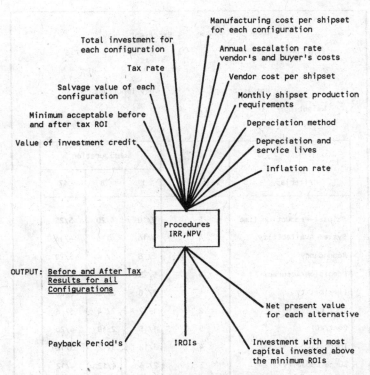

Total investment for
each configuration

Tax rate

Salvage value of each
configuration

Minimum acceptable before
and after tax ROI

Value of investment credit

Manufacturing cost per shipset
for each configuration

Annual escalation rate
vendor's and buyer's costs

Vendor cost per shipset

Monthly shipset production
requirements

Depreciation method

Depreciation and
service lives

Inflation rate

Procedures
IRR, NPV

OUTPUT: Before and After Tax
Results for all
Configurations

Net present value
for each alternative

Payback Period's

IROIs

Investment with most
capital invested above
the minimum ROIs

Figure 5-11. Input Data to the Investment Analysis Procedures, NPV and IRR

criteria, and the number next to each criterion is the weighting factor mutually agreed upon for the study. The diagonal in each box on the right-hand side separates the rating of the configuration with respect to a criterion from the "score" for that criterion. The rating is a relative value from one to five conveying the relative compliance of the configuration with that criterion (five indicating that the configuration completely satisfies that criterion) and the "score" equals the rating multiplied by the weighting value for that criterion.

The matrix was useful in evaluating each configuration with respect to intangibles—criteria that are difficult or impossible to evaluate strictly mathematically. Criteria such as flexibility, accuracy, redundancy, and so on, are typical intangibles.

Configuration 17 had the highest total score. It was chosen as the configuration on which to base the RFP.

```
          Poor      ---->      Excellent
Ranking:    1    2    3    4    5


          Not Important  ---->  Very Important
Weighting:  1    2    3    4    5
```

		Configuration		
Criteria	Weighting	1	8	17
Shipset Production Time	5	2/10	4/20	5/25
System Availability	4	4/16	4/16	4/16
Redundancy	4	2/ 8	1/ 4	5/20
Precision/Accuracy	5	5/25	4/20	4/20
Flexibility	4	2/ 8	2/ 8	5/20
Tool Capacity	2	1/ 2	2/ 4	3/ 6
Cost/ROI	5	1/ 5	2/10	4/20
Inspection	4	3/12	3/12	3/12
Surge Capacity	3	2/ 6	4/12	4/12
Total Score:				151

Figure 5-12.   General Electric Configuration Evaluation Matrix (Subset of Total)

Final Choice of a CIM System Configuration

Completion of the systematic configuration evaluation procedure described here will indicate which of the CIM design configurations, if any, should be chosen to produce a given group or family of parts.

The word "indicate" is crucial here, for the evaluation cannot choose the "best" configuration. The final decision must be tempered by judgment. Alternatives to a CIM system must also be considered.

Chapter 6

CIM Procurement

After choosing a CIM system configuration design that best satisfies the evaluation criteria, it is necessary to convey the findings of the study to potential vendors who can provide the system. This is accomplished by creating a procurement request for proposal (RFP), which is step 4 of the implementation sequence. Figures 6-1, 6-2 summarize this step.

The elements of a typical RFP depend upon the buyer's experience and willingness to become involved in the design process, as well as his desire to control the specification process and performance of CIM after it is installed and operating. At a minimum, the vendor needs:

• Drawings of the typical parts to be produced;
• Yearly production requirements and available production time;
• System delivery date.

However, this minimal specification approach should be avoided. If the buyer has completed the design and evaluation steps, he will be able to write a more complete RFP and will be in a good position to work with the vendors and to evaluate their proposals.

Strategy for Writing a Request-for-Proposal

In reviewing the myriad of topics that could be included in an RFP, it is important not to lose sight of its overall goal: to obtain quotes from a number of vendors, each of which may have a different approach. Also, although it is possible to reduce the number of request/proposal iterations by considering details in advance, it is unlikely that the final vendor will be chosen in response to the first RFP. CIM systems are relatively new and quite complex; neither buyers nor vendors have had

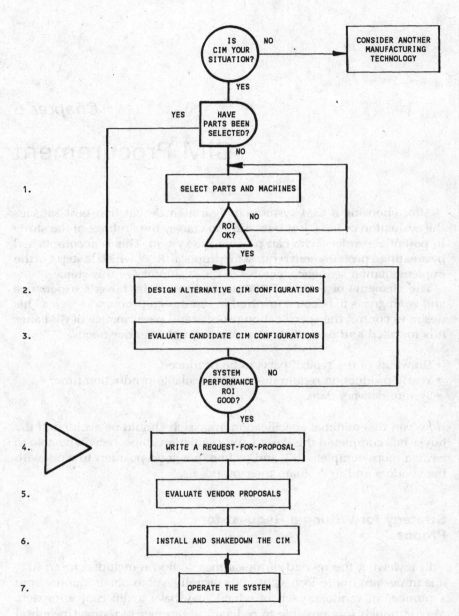

Figure 6-1.   Acquisition of A CIM System: STEP 4

- Write an RFP that conveys your findings and desires for a CIM system.
- Avoid overspecification; allow CIM vendors to be creative and competitive in designing for your situation.
- Data required for writing the RFP:
  1. Necessary:
     — Part drawings.
     — Annual production requirements.
     — Available production time.
     — Accuracy requirements.
     — Performance tests expected.
     — Desired system availability.
  2. Recommended:
     — Process plans and fixturing concepts.
     — Desired auxiliary equipment, such as inspection machines, part wash stations, etc.
     — Desired software capability.
     — Available floor space.
     — Utilities available.
     — Desired system redundancy.
     — Delivery date.
  3. Optional:
     — Capacity of the system, in terms of maximum machining horsepower, tool storage, maximum part dimensions (machining cube) and weight, etc.
     — Equipment biases.
     — MHS components desired.
     — Plans for future expansion.

Figure 6-2. Step 4: Write a Request-for-Proposal (RFP)

enough experience producing and operating them to be able to specify a system definitively in one iteration. There may be three or four iterations between the initial request and the signing of a contract. In between, the specifications become more detailed and well defined, and the number of prospective vendors decreases.

There is a strategy, therefore, in what to specify in the RFP and what is best left unspecified. In this way, the number of potential vendors is not limited nor their creativity stifled, but the buyer still controls the overall system design.

The major consideration, then, is the amount of information to include in the initial RFP. This can range from specifying only the production requirements and supplying the part prints to specifying the entire system (machines, MHSs, tools, layouts, etc.).

It is desirable, when possible, to include process plans and the desired fixturing approach for the part set with the initial RFP. There are three reasons for this:

1. It greatly reduces the amount of work a potential vendor must do and encourages more vendors to bid.
2. It assures that parts will be machined in an acceptable fashion.
3. It provides an equal basis for comparison of various proposals.

This is not to say that the vendor may not deviate from these plans; if a better fixturing concept or sequence can be developed, he should be allowed to examine it.

If the buyer is not sure about the need for certain equipment, such as an automated MHS or automatic inspection machine, they should be left as "options" for the vendors to justify if they choose to include them in their system design.

A final consideration is in how to present the RFP to the vendor. The most common approach is to mail each potential vendor (determined by the buyer's staff) the RFP packet and wait for a reply. Alternatively, the vendors can be invited to a bidders' conference at the buyer's facility to receive a copy of the RFP and initiate an exchange of ideas. This way, it is possible to form an early impression of the vendors' capabilities.

Elements of the Request-for-Proposal

Specifications

Specifications for a CIM system can be categorized into three major groups: mission specifications, performance specifications, and subsystem specifications.

Mission specifications include:

• Drawings of the parts to be produced;
• Process plans and fixturing concepts;
• Yearly production requirements and available production time (alternatively, system throughput), plus surge capacity required;
• Delivery date.

Performance specifications include:

- System availability (as opposed to stand-alone machine availability);
- Desired redundancy;
- Accuracy requirements;
- Performance tests (qualifying and acceptance tests) for individual machines and for the entire system.

Subsystem specifications include:

- Physical "capacity" of the system, such as horsepower, tool storage capacity, maximum part dimensions and weight, computer power, etc.;
- Machinery desired, such as horizontal or vertical machining centers, dedicated drilling/tapping machines, head changers, etc.;
- Auxiliary equipment, such as automatic inspection machines, deburring stations, heat treating equipment, washing stations;
- Desired software capabilities, e.g., details of management information system, diagnostic packages, decision support packages, scheduling and dispatching packages, modularity, a system simulation, etc.;
- MHS components (including pallets and fixtures) and topology (including available floor space and utilities);
- If necessary, standardized controllers. (This may rule out certain vendors or may reduce certain system capabilities.)
- Applicable industry standards, such as those for NC part programs and languages and the needed computer interfaces, such as IGES/PDES and MAP/TOP;
- CIM operations manuals;
- CIM software documentation.

Those specifications not stated will be filled in by the vendor. The level of specification detail will determine how tightly the designs are controlled.

Mission Specifications

The most critical of the specifications provided are the part drawings and the required system production requirements.

The part drawings should be as clear and detailed as possible, and they should represent the latest revisions. As an aid to the vendor, critical or unusual tolerance requirements should be highlighted in some way that attracts attention. Including more than one set of drawings allows the vendor's various systems groups to begin work on a proposal at the same time.

As an additional aid for the vendor, as well as providing a means for an equal comparison between vendor proposals and assuring that the parts can be made, the vendor should be provided with process plans and fixturing concepts for each part.

System production specifications include the production requirements of the system (stated as either a monthly or yearly figure) and the allowable working hours that will be available for system operation. In addition, if there will be an occasional need to exceed the average throughput specified, a "surge" capacity should be stated. Finally, specify the desired flexibility to changes in the parts or part mix that may occur during the system's life.

A key element in a vendor's proposal is the delivery schedule. If this is to be specified by the buyer, the dates should be as realistic as possible. System purchases are not the same as single machine tool purchases. The amount of lead time required to design, fabricate, test, ship, and install a system is different for each vendor and for each system design. The lead time for many first system operations is in excess of 12 months from the time a purchase agreement is signed. The shakedown period required to bring the CIM system up to full production is an additional six months or more.

Performance Specifications

The desired level of system availability—the percentage of time all components of the system are functioning normally—can be stated in the RFP, but this value is very difficult for either the vendor or the buyer to estimate accurately. The vendors will have information gathered over time on the availability of systems they have previously installed, and this figure averages from 65 to 85 percent. Stating a desired availability of 75 percent will indicate to the vendors the minimum amount of availability that will be acceptable in normal operation, and they should provide enough redundancy in the system to provide that availability.

Redundancy is the amount of "backup" in the system for each subsystem. For instance, if the system contains at least two of every type of machine and fixture, production can continue while a failure is repaired. The more of any one machine there is, the less likely a failure of one is going to affect production. However, redundancy is expensive; a balance should be struck between redundancy and cost. It is possible to examine the cost of redundancy using economic analysis to estimate the cost of not producing an item or producing it late and comparing to the cost of having redundant equipment which may not be used fully.

If there are unusual accuracy requirements, they should be highlighted and the vendors should be expected to discuss their solutions to these problems in detail. This discussion might include: the best accuracy the vendor feels he can obtain in normal practice, the option of taking unusual precautions (with accompanying cost increases) in building the machines to obtain higher intrinsic accuracy, how he might use manual intervention to inspect and reset boring bars, the use of probes in the machine spindle, etc.

Performance tests usually include results of detailed simulations of the proposed system plus documented proof of the accuracy and capability of each machine. The buyer should request that the vendor, using computer simulation, verify that his proposed CIM system configuration will achieve the necessary average throughput desired. The simulation should also indicate the utilization of each of the system elements and provide evidence as to the surge capacity. The effects of machine and MHS failures on throughput should be examined at different failure rates to assure that the desired redundancy has been built into the system. The buyer may also ask that various batching, balancing, scheduling, and dispatching strategies be simulated. This will indicate which are best and will document the effects of each. The results of these simulation runs should be provided for review either with the proposal or at the final specification stage.

Performance specifications for each subsystem element can also be included in the RFP. These tests are most often used to verify the integrity and inherent accuracy of the machines. They usually involve tracing a master calibration piece with a measuring probe and examining the error between the actual and programmed movement of the machine for various temperatures and lengths of time. (After system installation, it may be wise to permanently mount such a "master gauge" to a pallet for periodic inspection of machine alignment, and for the analysis of the effects machine "crashes.") Machining tests (qualifying tests) may also be specified, but the results of this type of testing procedure are much more susceptible to the introduction of errors from sources other than the machine, such as the pallet, fixture, part, cutting tools, etc. The vendor should be required to document the results of all tests for all machines, so that if a problem with a machine occurs at some later date, both the buyer and the vendor have a common reference point from which to start their investigation.

Information on the desired system and subsystem acceptance tests, to be performed both at the vendor's site before shipping and at the buyer's plant after installation, should be included in the RFP in detail, to prevent misunderstandings later. These tests usually require a number of actual production pieces to be run off. The number of pieces to be machined in each case and the limits of acceptable error are usually agreed upon after a vendor presents his proposal and it is accepted, but it is important to provide enough information for the vendor to estimate the cost and time involved.

Subsystem Specifications

Subsystem specifications include optional information on specific attributes of the equipment the vendor will provide. This can include the physical capacity of the system, such as machine horsepower, pallet sizes, machine travel, computer memory size, number of material handling units, and so on. Any software capacity constraints should also be mentioned. This information should be given if the CIM system might be used in the future for parts which differ in size, material, or weight from the proposed part set.

Specific types of machines desired should be mentioned if there is a preference. For example, horizontal machines may be preferred because the existing tooling is designed for horizontal machines, the process planners are familiar with horizontal machines, etc. It may be requested that the vendor investigate some dedicated equipment, such as head-changers, cluster heads, etc., where it is believed the equipment might be cost effective. However, it is best not to stifle the vendor's creativity: provide guidelines, not rules.

Preference, if any, for certain types of MHSs should be mentioned in this section. If the floors are uneven, shallow, or poorly surfaced, with small turning areas, or the machines will be arranged in an unusual fashion, the vendors will need this information to choose their MHSs. In addition, specify possible future expansion requirements for the line.

Auxiliary equipment—such as automated inspection equipment, deburring, etc.—are usually best treated as options, due to the fact that not all vendors can supply all of these items and because the justification for each item may be desired separately.

Additional specifications that should be provided to the vendor include those of available floor space, atmospheric control, and utilities available (heat, power, light, chilled water, compressed air, etc.). Interface specifications are necessary if the system is to communicate with equipment already installed. Also, specify any applicable standards, such as those for NC part programming languages. Last, inform the vendors of the desired proposal format—how to divide the system into components and price those components—so that the evaluation is not hindered.

System Control and Monitoring

Overall control is important because just as much as the machines, it determines how well the system will function. It is centralized in a computer that has direct data and system status links to the marketing, design, processing, material handling, tool management, inspection, and delivery functions.

The architecture of the control defines the relationship between the computer and the other system elements, and it can vary significantly between CIM manufacturers. Modularity, subsystem autonomy, reconfigurability after machine failure, and growth and operational expansion all depend, in different degrees, on how the software system is structured.

Operation of CIM involves, in its simplest form, the following functions:

- Part program preparation;
- Part program loading;
- Machine scheduling and operation;
- System monitoring;
- System diagnostics;
- Operator displays and controls;
- Tool management

The CIM system is likely to operate in several modes and the control system must adapt to each. There is automatic or full operation, where all systems are working as desired. There is a setup or changeover condition for a short-term, partial failure mode where the system may be operated in a semiautomatic fashion. Also, there is a full failure mode, where a severe failure precludes operating the components as a system. The control system determines how to obtain the best response in all these situations; it is important to pay attention to it early in the specification process. By specifying that the system degrade "gracefully," backup features that allow the CIM system to function to some degree when individual subsystems fail should be included.

Although standard software is provided, additional software may be desired. A management decision support system is often needed. Such a system might include batching/balancing software, real-time scheduling routines, a simulation for examining changes due to failures, and tooling status displays.

Another area of special software includes line monitoring and diagnostics. Timely detection and identification of failures greatly reduces the effort required to service the CIM system, since failure diagnosis usually demands much more time than correcting the fault.

The vendor may also be requested to develop NC part programs in the language the buyer currently uses, as well as process plans or operation sheets. A very useful software feature is the ability to edit and update programs either at individual machines or from the CIM control computer. This saves time and allows optimization in real-time for the machines that will make the parts. NC tape verification, using a plotter or cathode ray tube (CRT) graphics package, should also be considered.

Documentation

Documentation for all equipment must be provided to include operation manuals, software manuals, maintenance manuals, and recommended spare parts lists. Multiple copies are recommended, particularly of the software itself. Also, specify the frequency of contract progress reporting as well as the principal items to be covered in the reports.

Vendor Responsibility

In the case of a system that includes machines from more than one vendor, there is the question of who will take responsibility for properly integrating the equipment. Normally, buyers do not have the expertise to perform this function. In general, if inspection machines, high accuracy machines, etc., from other vendors are to be integrated into the line, the overall responsibility must be placed on the shoulders of the prime vendor. If there is to be a service team at the plant during installation, include that also. Finally, describe the amount of subcontracting the vendor will be allowed and how that subcontracting is to be accomplished.

The preparation of the site, machine and MHS foundations, service and utility pits, coolant troughs, etc., are almost always the buyer's responsibility. Integrity of the foundations plays a very large role in smooth system operation.

Post-Installation Support

Two important areas of post-installation support that should be mentioned specifically in the RFP are the system's warranty period and the services to be provided after installation. Both topics are negotiable. The warranty usually covers materials and workmanship of all system elements and software for a minimum period of one year. If an element fails within the warranty period, the warranty pertaining to it is often extended for a year from the date of repair. Post-installation service might include periodic inspection of equipment, supply of spare parts, and links to a diagnostic computer at the vendor's plant. Clauses for updating obsolete hardware or software should be made clear.

Additional System Options

Inspection

Available inspection methods include manual inspection with standard instruments; using computer coordinate measuring machines; and inspection with new techniques such as machine vision, optical imaging, laser interferometry, and other sensors. How these techniques are used, not necessarily which ones are used, can make the difference in a CIM system.

Inspection philosophies include:

• Preprocess inspection on the machine tool, to verify head alignment, spindle concentricity, tool position, etc.
• In-process checks for dimensional control, detection of tool wear, plus adaptive control of cutting speed and feed rate. Included in this philosophy is the use of spindle probes for part alignment (determining axis compensation) and part and feature presence (hole or no hole, excess stock or lack of stock).
• Post-process inspection, to verify and document the correct dimensions and finish.

No matter what the techniques and philosophy, six questions with respect to specifying a system must be answered:

1. Should inspection be on-line or off-line?
2. How many inspectors or inspection stations are needed to maintain the desired production rate?
3. Should there be statistical sampling or 100 percent inspection? What features should be inspected?
4. What inspection data would be most useful to the control system; how much data will be archived?
5. Should parts be deburred and cleaned before inspection? (If so, additional stations for deburring and cleaning will have to be added.)
6. Should parts be unfixtured or fixtured for inspection?

The current trend is to on-line coordinate measuring machines which communicate with machining center controls and offset some machining errors. They can monitor quality in real-time and alert the system operator to problems before a significant number of parts are affected. These machines are costly and fairly slow; more than one may be necessary to maintain throughput.

Measuring probes, stored in the machine tool changer and interchanged with tools in a machine's spindle, are also popular, but they reduce the time the machine potentially could be cutting metal.

Chip and Coolant Recovery

The major issues involving chip and coolant recovery focus on the question of a centralized chip- and coolant-recovery facility or individual chip conveyors and coolant systems on each machine. The advantages of collecting chips automatically in one spot and the ability to monitor directly the properties of the coolant temperature, water content, foaming, etc., should be considered. Additionally, a wash station for the parts/fixtures/pallets before inspection is easily facilitated with a centralized system, and coolant could be used to wash down the system at the end of each shift.

However, individual chip- and coolant-recovery units on each machine minimize the machine dependency on a centralized system, allow simultaneous cutting and recovery of different materials on different machines in the system, and do not require the site work and space needed by a centralized system.

Cutting-Tool Room (Tool Crib)

CIM tools can be serviced from an existing tool crib or from a dedicated CIM tool crib. If a dedicated tool crib is chosen, a decision must be made as to the level of complexity that might be appropriate. In general, dedicated CIM tool cribs have proven to be cost effective for a number of reasons:

• The tool crib is under the control of the CIM manager, simplifying communications and supervision.
• The tool setters become familiar with the CIM tools and tool life.
• Tool management is improved.
• Response to tool failures or missing tools is rapid.
• Process or part changes are accommodated more quickly.

This is not to say that an existing tool crib cannot be used; if it is well managed and capable of handling the increased demand of an automated system, then it might be the best choice.

The complexity of a dedicated tool crib can vary widely, from having simple, manual presetting and sharpening machines to having automated presetting and sharpening machines under computer control that send tool length and diameter information to the machine tool on which the tool will be used. The buyer must work with the vendor to determine the appropriate level of sophistication.

Decide whether the vendor will provide the perishable and durable tooling, but reserve the right to change specifications or purchase unacceptable proposed tooling elsewhere. Tool crib equipment and tool setting equipment should be described; a tool identification system and pallet specifications must be mentioned. The buyer, the CIM vendor, and the raw materials supplier will have to discuss and mutually agree upon the final configuration of the fixtures.

Unmanned Operation

If the intention is to operate the system in an "unmanned" mode for one or more shifts a day, a number of options should be considered:

- Video monitoring of each station from a central control area.
- Various means of adaptive control at each machine, such as spindle- or feed-drive-motor current monitors.
- Tool breakage and wear monitoring at each machine.
- Duplicate tooling at each machine.
- Reduction of feeds and speeds.
- In-process inspection as well as an inspection machine.
- Extra pallet/fixture storage area for preloaded fixtures.

As with the tool room, the buyer and the vendor must determine the degree of sophistication which is both possible and advisable.

Evaluating Vendor Proposals

One or two months after issuing the RFP, budgetary proposals should have been received from all of the interested vendors. As each vendor's proposal is examined, make entries in a decision matrix indicating the relative score of the proposal for each major topic. The proposals can then be reviewed as a group.

This section describes Step 5 of the CIM implementation sequence and is highlighted in Figures 6-3, 6-4.

Review of Individual Proposals

A simple subjective rating scheme—numbers from one to five—can be used to evaluate the desirability of each proposal with respect to each evaluation criterion. Each criterion should be given a relative weighting, to indicate which are most important. The matrix is useful for comparing the group of proposals, since the buyer can simply scan a line and ascertain how well a proposal addresses each topic or how well one proposal fares against another.

Although they are difficult to separate during the evaluation process three areas concern buyers the most:

1. The cost of each element of the system;
2. Performance verification of each element as well as of the system itself;
3. The reputation and experience of the vendor.

Cost should not be the only criterion for choosing a system. Each element of the RFP quoted on must be examined thoroughly to determine exactly what it includes. Questions to consider when rating each element with respect to cost are:

* Are engineering costs included clearly in the proposal?
* Are site work costs included?
* What types of support (during and after installation) is the vendor offering at the quoted price?
* Is a training program offered?
* What spare parts are included?
* Can a recommended spare parts package be purchased?
* Is an installation team included in the quote?
* Are all specified elements included, except options?
* Has the system been over-designed? Under-designed?
* Are the costs reasonable with respect to the delivery date and equipment capability?
* What are the operating costs—manpower, utilities, and off-line operations—in this proposal?
* Does the vendor have local service capability?
* Can the system be installed in phases?
* Is the system expandable?

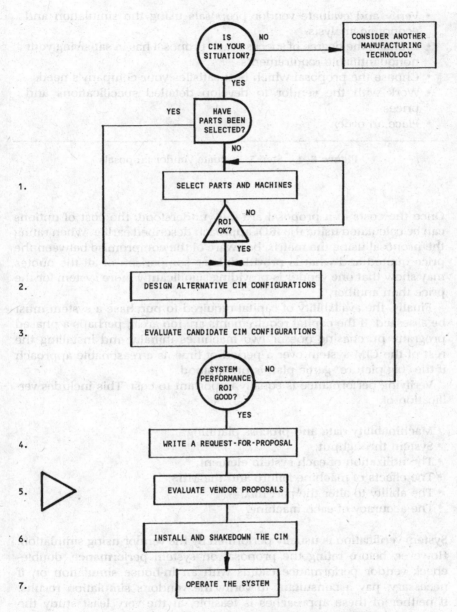

Figure 6-3. Acquisition of A CIM System: STEP 5

- Verify and evaluate vendor proposals using the simulation and economic analysis.
- Evaluate the degree of success each proposal has in satisfying your nonquantifiable requirements.
- Choose the proposal which best satisfies your company's needs.
- Work with the vendor to develop detailed specifications and prices.
- Place an order.

Figure 6-4.   Step 5: Evaluate Vendor Proposals

Once the costs of a proposal are fully understood, the cost of options can be calculated using the IROI approach described earlier. When rating the proposal using the matrix, be aware of the compromise between the price quoted and what is provided; later comparison of all the quotes may show that one vendor is providing significantly more system for the price than another.

Finally, the availability of capital required to purchase a system must be assessed. If the capital requirements are too great, perhaps a phased program, purchasing one or two machines initially and installing the rest of the CIM system over a period of time, is a reasonable approach if the "big picture" game plan is understood.

Verifying performance is equally important to cost. This includes verification of:

- Machinability data and process planning;
- System throughput;
- The utilization of each system element;
- The effects of machine failure and the MHS;
- The ability to alter the part mix;
- The accuracy of each machine.

System verification is usually performed by the vendor using simulation. However, before rating the proposal on system performance, double-check vendor performance claims with an in-house simulation or, if necessary, pay a consultant to verify the vendor's simulation results. If neither of these approaches is feasible, at the very least study the vendor's simulation to make certain that it is modeling the proposed CIM system accurately and completely.

Verifying the performance of the vendor's individual machines is more difficult. If possible, the best approach is to talk to another user of the same machines. This is usually much more realistic than the vendor's estimates of accuracy, since the machine in the shop has had a chance to wear in and sustain normal production punishment.

Do not underestimate the usefulness of vendor reputation as a final guide. It can indicate whether problems in delivery are likely to occur, as well as the likely quality of the long-term support. Personal experience with the vendor's stand-alone machines also should play a part in the evaluation. Remember, however, that a separate division of that vendor's company may be offering the CIM system and that division may have an entirely different set of policies.

Selection of the Vendor

After completing the proposal evaluation matrix for all of the potential vendors, the vendor with the highest total score can be selected, and then, contract negotiations begun.

Final Procurement Specifications

After selecting the vendor, the final RFP must be converted into a final specification for the CIM system. All parties involved at the buyer's plant should have input to this document and agree to any changes. This specification should be the negotiating vehicle to obtain a final agreement with the vendor. The document must state precisely the content and capabilities of the system, as well as the responsibilities of all of the parties. Upon reaching a final agreement on the specifications, a purchase order can be issued for the system.

Chapter 7

Installation of CIM

The buyer's involvement with CIM implementation does not end with signing the purchase agreement. Although the vendor is supplying all of the parts and will help the buyer put them together, the buyer must prepare for CIM, assist the vendor in its installation, and begin to debug the system. Every effort made to maintain a partnership with the vendor during this phase of the project will provide dividends in the long run.

This section outlines Step 6 of the CIM implementation sequence, summarized in Figures 7-1, Figure 7-2.

Preparing to Take Delivery of a CIM System

Labor

CIM systems are a relatively new approach to manufacturing and can result in a significant change in many plant operations. As such, they affect people on all levels of responsibility, from the shop workers and supervisors up through the ranks of management. CIM will be regarded with different perceptions, ranging from a grave threat to a remarkable opportunity.

Do not overlook the necessity to remove, as much as possible, negative reactions to the installation. The threat of job displacement or job elimination is most obvious. In addition, new technology often appears overwhelmingly complicated to those who have not been schooled in the basics. Thus, it often has a psychological effect of diminishing the self-confidence that motivates skilled workers and supervisors.

CIM can present an opportunity to generate new skills in employees, and it should be used as such to enhance career possibilities. It can be a motivational source to develop new skills for workers whose previous

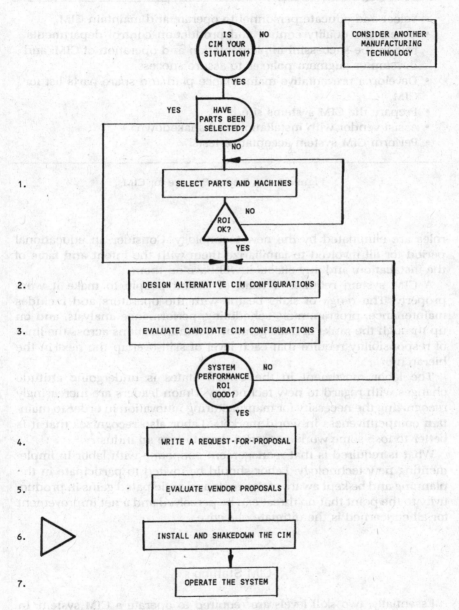

Figure 7-1. Acquisition of A CIM System: STEP 6

- Select and educate personnel to operate and maintain CIM.
- Assess the quality control and production control departments' role in the successful implementation and operation of CIM, and develop or augment policies to assure success.
- Develop a preventative maintenance plan and spare parts list for CIM.
- Prepare the CIM systems site.
- Assist vendor with installation and shakedown.
- Perform CIM system acceptance tests.

Figure 7-2. Step 6: Prepare for CIM

roles are eliminated by the new technology. Consider an educational period for all involved to familiarize them with the intent and facts of the installation, and any effects it will have on them.

A CIM system requires a team of skilled people to make it work properly. The range of skills begins with the operators and includes maintenance, programming, scheduling, performance analysts, and on up through the ranks of management. Communications across the lines of responsibility require that each level of skill overlap the next in the hierarchy.

The labor movement in the United States is undergoing attitude changes with regard to new technology. Union leaders are increasingly recognizing the necessity for manufacturing innovation in order to maintain competitiveness in world markets. Labor also recognizes that it is better to lose some work categories than to lose an industry.

What is required is that management cooperate with labor in implementing new technology. Labor should be invited to participate in the planning and be kept aware of the goals and anticipated gains in productivity to the point that no threat can be perceived and a net improvement for all concerned is the ultimate objective.

CIM Staffing

Essentially, two skill levels are required to operate a CIM system. In the unskilled classification, there is the loader/unloader. The most demanding elements of his job entail making sure that part and fixture mounting surfaces are clean so that parts register properly, and sometimes clamping parts at prescribed torque levels to prevent excessive part strain.

In general, highly skilled machine operators attend machines during the shakedown of a CIM system. They are a logical choice for start-up because they can readily identify problem areas associated with the machine tools. Often, they will not physically operate the equipment, but monitor it to determine whether its operation is satisfactory. Their value continues beyond this initial phase, because it is very important to keep the machines running, but they are treated more like consultants, called only when a problem arises that is too difficult for the line foreman to take care of.

In the skilled classification, there is the line foreman, typically a manufacturing engineer with machining and systems engineering background. He is responsible for maintaining production on the line and must make decisions affecting raw part orders, manpower placement when failures (machines, computer, controllers) occur, and perhaps even part routing in the event of such failures.

The number of loaders depends on the number of load/unload stations and the part cycle times. A loader should be able to work several stations; the exact number requires a time study for the particular CIM system in question. Similarly, the number of machine operators required will depend on the particular line. One operator per machine is not required, as the "operator" does not normally operate any machine, unlike common practice for stand-alone machines. Under normal circumstances, one worker per five machines may be adequate.

Some users have found that there are significant benefits to be gained when most personnel running the system have the skills needed to do the other jobs as well, from fixturing parts to supervising the system. Aside from the obvious advantage that absentees can easily be covered by others, users find that surges can be handled better. For example, there may be a sudden need to load eight parts on their fixtures. It is very helpful if all "hands" can respond to the need. In addition, this is a method to provide job enrichment, especially if the CIM team is allowed to assign their work on a revolving basis by group agreement.

The ability to operate in this manner depends strongly on union policies. It may require that the same job description, pay, and ranking, etc., be given to everyone.

Quality Control

CIM can be an unattended system (as in Japan), and there will be no operators to visually inspect the parts for missing holes, cracks, or other material defects. Nor will the NC programs accommodate large casting variations as a skilled machinist can. So there is a need to adequately evaluate the casting vendor, and incoming inspection becomes even more important.

One company, expecting its CIM system to be installed "turnkey," relied on its old approach: If a casting chosen at random once a year passed a layout inspection, the vendor was accepted, and castings were brought into the factory with nothing more for incoming inspection than an occasional visual check. This policy had been satisfactory when skilled machinists could be counted on to accommodate excess stock and check for missing cored holes and other defects. The CIM system, however, had no way of inspecting the parts, and machine crashes and scrapped parts set production back more than one year. Confidence in the system was destroyed, and only after a detailed examination did the company realize that its inspection policy, and not the CIM system, needed improvement.

The quality control and production control personnel will need to decide whether rework, part program proofing, and fixture/tool verification would better be performed on-line or off-line with duplicate machines. For CIM systems, the decision to go off-line is usually preferred. There are a number of reasons:

• Reintroducing the part back into the system may be disruptive.
• The tooling required may no longer reside at the machine.
• There is usually only one feature out of tolerance; running the part completely through the system would be a waste of time and rescheduling the part on one machine would not be worthwhile.
• Off-line rework equipment can be pallet-compatible with CIM. (Additional pallets and fixtures may be needed.)

Determining the source of an "out of tolerance" problem in a CIM system is often not easy. The error may be a result of any one or a combination of factors, for example, tool wear, interface misalignments (tool/spindle, part/fixture, fixture/pallet, environment, pallet/machine), etc. CIM systems that incorporate redundant tooling present an additional problem. The computer must keep track of which tools on which machines were used on which workpieces. If inspection is done on a sample basis, workpieces traveling different routes should be sampled frequently.

Production and Inventory Control

Lot-size calculations are important because they give some idea of how much support CIM will require from inventory. It is important to estimate the number of casting producers necessary, warehouse space, and time for material handling.

Lot sizing is a twofold problem. First, determine the optimum production lot size based on:

• Machinability and tool life;
• Set-up costs;
• Scheduling requirements;
• Planned system capacity;
• Average machining time;
• In-process inventory carrying costs.

Second, estimate the optimum inventory lot size, considering:

• Warehouse space;
• Carrying costs;
• Ordering costs;
• Demand requirements.

Calculate inventory lot sizes twice—once for raw castings and once for finished castings—to determine total space requirements.

An in-place manufacturing resource planning system (MRPS), or some other kind of scheduling system, will be greatly influenced by the introduction of CIM.

Preparations for Maintenance

Because of the CIM system's complexity compared to stand-alone machines, maintenance planning is a necessity. A competent CIM maintenance organization should be created when a CIM system is contracted, not after it is delivered. Every effort to contact and work with other current users of CIM systems from the manufacturer will pay great dividends.

Each of the many disciplines of CIM must be addressed in creating a maintenance force. These include specialists in computer hardware and software, controls, machine tools, tooling, and manufacturing. A strong emphasis should be placed on programming and computer operations because CIM software will be new to the organization. CIM systems have not been turnkey installations; users have historically experienced long learning curves due to underestimating the maintenance skills necessary.

Development of a spare parts list should be pursued with the CIM vendor at contract time, and it is useful to discuss it with users of similar CIM systems. The vendor is in a delicate situation. On one hand, he is trying to sell the reliability of his system; on the other, the buyer will need the spare parts and the vendor will profit from their sale. It is natural to delay investing in spare parts because the total cost can be considerable. But the lead time for these parts may be large, and ultimately many of them will be needed.

Installation Preparations

On-Site Preparation

Foundation design for CIM differs from those of stand-alone machines by the sheer extent of the system and by the potential need to integrate coolant and chip handling subsystems over the full set of machines.

Foundations should be designed to achieve excellent stability. Since the MHS interconnect machines and have some nominal component positioning accuracy requirements in order to function, there are greater demands on the stability of the system as a whole than there are for stand-alone machines. The foundations should be installed long before delivery of the CIM system to allow them to settle and stabilize.

An often forgotten issue at the design stage is the accessibility of the equipment for preventive maintenance and repair. Especially in automated system, thought should be given to providing sufficient access space around the machines and MHS. Consider these points when developing configuration layouts and, for example, consider MHSs that also could be run manually.

Power requirements for CIM are considerable when machine tools, MHSs, and computers are grouped. CIM components may be susceptible to moderate voltage variations, and CIM may affect other equipment on the same supply bus. Computers are particularly sensitive to power conditions and warrant sound energy supply practices.

Off-Site Preparation

After a vendor is chosen, form a liaison team to work with the vendor to fine tune the system design and gain assurance that the system will do the job intended. The team will aid in developing the physical interface with the manufacturing plant and oversee installation preparations.

If the CIM system is a multivendor design, the team can make certain that the vendors are working together. (Usually this is not a problem since one of the vendors should have prime responsibility.)

Other Preparation

In order to become familiar with the dynamics of CIM operation, exercise the computer simulation model of the system, investigating part changes, part-mix changes, and machine and MHS failures.

Presumably, at least a tentative part mix and its attendant tool complement will be chosen long before the CIM system is installed. This in itself serves to define tool room requirements. If the machinability of the parts is low, extra tooling should be stocked to offset anticipated breakage.

If during CIM construction you change parts or part mix, the changes should be finalized before the system is delivered. While a CIM system is progressing to a fully productive state, keep other operating factors stable.

An effective training program must be used to introduce CIM to the operating and maintenance staff. The vendor may provide all of this service or just parts of it. Its extent should be defined in the original contract.

Installation and Shakedown

Machine and System Acceptance Tests

There should be at least two levels of acceptance tests called out in the RFP and subsequently agreed to in the contract. The first measures the performance of machines and perhaps other subsystems on a stand-alone basis. The second is the acceptance test for the entire system.

Stand-alone machine acceptance tests can take two forms. The most common one, known as a "qualifying" test, demonstrates the ability to manufacture parts to production accuracy specifications. The test pieces may not be actual production parts, but are representative in shape, material, and machining operations. They usually are simpler than production parts and might take the form of a boxlike weldment; they incorporate all the significant operations associated with production parts, such as face and pocket milling, drilling, tapping, boring, and turning.

Test cuts directly show the performance of a machine tool in producing a tangible product. However, the machine's inherent accuracy may be partially masked by sharp new tools, nonoptimum process plans, etc. The second common acceptance method is known as the master part trace or "performance" test. A fully machined and inspected part is clamped on the pallet and traced with a probe using a special NC program for that part. Probe deflections are a direct measure of machine positioning errors. This technique presumes finish-cut machining does not impose significant loads on the part and, therefore, machining errors essentially are the result of positioning errors. This method will, therefore, not measure errors due to lack of machine rigidity, lack of part/fixture rigidity, etc.

The complete system acceptance test is a demonstration of the full system functioning in production mode producing an agreed-upon number of parts within a specified time. Sometimes the entire system—or perhaps each individual component—must remain free of all failures for some minimum percentage of the time.

Typical Shakedown Problems

Every installation is plagued with start-up problems. Most CIM systems are custom designed and built; although the machines and MHSs are usually standard off the shelf items, many of the problems are completely new with each installation.

In multivendor systems, lead times for machine delivery may be quite different, and only a portion of the CIM system may be installed for some time. This can be an opportunity to learn the operation of and check out certain elements of the system under less pressure. However, if some key elements are missing, it may not be possible to do anything but treat machines as stand-alones until the full system is integrated.

Software "bugs" can range throughout the computer-controlled system. Some examples of the kinds of problems that occur are a sudden inability to read tapes despite having done so before, an incomplete control-software checkout that only reveals itself under certain operating conditions, conflicts in control logic, etc. The buyer's software people should work with the vendor's programmers to solve these problems.

MHS interface hardware may develop mechanical problems that were thought to be engineered out during system design. Once again, it must be emphasized that the uniqueness and complexity of the system promote growing pains that are not circumvented just by good design practices.

Normal manufacturing problems, like machinability, can be compounded in a CIM system because of the interdependence of machines. What was believed to be the line's pace-setting machines, may not be so after their speeds or feeds are adjusted to achieve adequate tool life.

Part changes should be minimized during shakedown. It is much easier to learn to operate a CIM system with a fixed set of parts than to have to change part programs, which may unbalance the line and require extensive changes in the part/machine mix. In this view, it also is better not to load a new CIM system to capacity during a start-up period. Checking out all the systems with just a few parts should be easier and will reveal major problem areas.

The lack of well-established maintenance schedules tends to be self-correcting in time as the needs of the system are assessed through experience. The supplier, of course, should recommend some basic schedule.

Diagnostic and management information systems aid in locating problems and recording performance measures during operation of the system. Both also can be extremely useful when debugging a system during installation and shakedown. The diagnostic system will pinpoint problem areas such as overlooked connections, mechanical malfunctioning, incorrect start-up or shut-down of the system, and so on. This information may be difficult to obtain otherwise during installation. The management information system (MIS) can continuously record performance statistics, providing a complete record of shakedown problems and progress.

Application Engineering

Engineering Activities

After the justification and approval to proceed with the installation of the CIM system, detailed engineering activities can begin. There are two engineering functions to be performed: application-oriented engineering functions and system/design engineering. Application engineering includes all of the steps leading up to the final specification and selection of the CIM system to be installed, the basic configuration of the workplace, and the details and supporting activities of the system.

Operations

The first step in application engineering is to review the operational aspects. For this review members from the CIM survey team should be used, including representatives from manufacturing engineering, industrial engineering, and plant engineering. Other members may be added as required, but the size of the group should be minimal to avoid interference and disruption of ongoing production operations. The group should monitor the operation to be automated during production times of all production shifts, during start-up and shut-down, and during continuous operation. The group should interview the production operators, material handling personnel, maintenance, and production supervisors to fully understand the operation and its support.

There are a number of considerations to be addressed. Each will have an impact upon the choice of equipment, the arrangement of the work cells, the process and material flow, and the human interface. An evaluation of alternatives and the course of action to be followed should be documented so that layouts, system engineering, equipment design, processing, scheduling, and training will all support the effective operation of the CIM system installation.

Data Collection

After the detailed operation review and evaluation, the group should begin collecting data about the operation, including:

• Number and description of elemental steps in the operation;
• Size, shape, weight, and so forth of parts or tools handled;
• Part orientation at delivery, acquisition, in-process, and at disposal;
• Methods and frequency of delivery and removal of parts;
• If the operation involves batch production—lot sizes, characteristics of all parts in family, frequency of changes, and changeover time for related equipment;
• Production requirements per hour, shift, day, and so on;
• Cycle times, both floor to floor and elemental;
• Inspection requirements and defect disposition.

Data sheets should be developed, each covering a different aspect of the operation and should all be combined into the final task, program, and equipment descriptions for the system.

When all the data sheets have been completed and reviewed, the group that performed the operational review, along with the manufacturing manager and any others having experience with automation technology or product expertise, should develop a detailed description of the intended operation. The task description should be detailed to the most elemental level possible—that is, motions of the machines or robots, each function of the tooling, each function of each related machine or device in the system, and all support functions, such as delivery and removal of parts—and should be described separately. The descriptions should include time elements, motion components (direction, distance, and speed for example), inputs, outputs, delays, and the like.

The task description will become the basis for a preliminary specification. Once this specification has been developed, the group should consider several systems that could meet it. Cost, delivery, reliability of the source, and other such considerations should not be addressed at this time; the intent is to pick out the systems most capable of performing the task and use them in the work area layout development. Layout development will reduce the number of selections. During the final specification development, economic and other issues will be addressed.

Work Layout

After all of the data about the operation have been collected, consolidated, reviewed, and some preliminary choices made, a layout drawing of the proposed installations should be prepared. Starting with a scale layout of the existing area, locations for incoming and outgoing material, buffers, and intermediate positions of parts, if necessary, are determined. From this layout, potential inference points can be located and equipment relocations, if any, can be developed. Sources and routing for utilities, such as electrical power, compressed air, and cooling water, if required, are also shown on the layout.

In most cases, floor plan layouts are sufficient to determine material flow, equipment arrangements, and potential interferences. However, where overhead conveyors are involved, or where the machines' task involves reaching into equipment or other restricted areas, elevation sketches should also be prepared to check access, determine reach requirements, and reveal possible interferences.

Selecting Optimum Configuration

Working with task descriptions and layouts, the group should now optimize the workplace arrangement and CIM operational specifications. The primary objectives are usually to minimize the number of program steps to attain the shortest CIM cycle time, the simplest programs, and the most reliable operations. The results of the optimization should be the determination of a final set of specifications, the reduction of choices to one basic configuration and not more than three or four possible approaches from which to choose, and the development of a practical work area layout.

Several alternative arrangements and operational sequences should be considered in the optimization process.

- Rearranging incoming and outgoing material locations or positions of the equipment can significantly affect the cycle time.
- Providing automatic material-handling systems that reduce the need to interrupt operations to replenish or remove parts can increase throughput.

- Using part feeders to provide initial orientation for parts' acquisition to eliminate an orientation step can save time.
- Product and/or process changes may also be necessary or desirable.

From these evaluations, the layout and machine combinations that appear to provide the lowest cycle time and/or the fewest motions per cycle should be selected. Other considerations include:

- Availability of more than one approach possible in meeting the requirements. Limiting one's choice to a single approach should be avoided for flexibility in cost, delivery, support, and so forth.
- Extent of rearrangements and/or relocations required. Extensive changes in workplace layout are costly, time consuming, and disruptive.
- Effectiveness in addressing all of the considerations reviewed during the first detailed study of the operation (backup, space, environment, and safety).

People Preparation

Up until this point in the CIM planning process, upper and middle management have been involved in the CIM process from its inception and are fully aware of its goals and status, but formal acknowledgement of the program and wide dissemination of information about it are uncommon. Now the preparation of people for the final phase of implementation will be equally important to the system integration.

These include production management, production supervision, skilled trade workers, direct labor and indirect labor employees. The interests, concerns, and expectations of each group differ and each must be approached with their specific needs in mind.

Production Management

Production managers' views of CIM may range from skeptical to hostile. They are generally concerned with output and with solutions to immediate problems. Lacking technological knowledge, they can be impulsive, impatient, and are more likely to be interested in the system's contributions to solving productivity, quality, and work force problems than in the expected economic gains.

Production managers must become involved in the final installation plan for the CIM system. They will be concerned about disruptions of schedules, loss of production, and start-up problems, and will be keenly interested in plans for providing backup for the system after start-up. Personnel safety is another element of concern, as are operator and maintenance training.

Production Supervision

Production supervisors are often caught in the middle between management and the work force. They may not be technologically oriented and are often harried and overworked. They will probably regard CIM as one more problem they will have to handle, rather than as a potential solution to some of their existing problems.

It is important, therefore, that production supervisors be provided detailed information about where and when the system will be installed and the company's plans for displaced workers. They should be provided with detailed information about backup plans and about plans for maintenance support for the system. Furthermore, production supervisors should participate in developing procedures for implementing those plans.

Skilled Trade Workers

Maintenance of CIM equipment is sometimes done by the supplier, by a third party on a contract basis, or by the company's own maintenance staff. When maintenance of the entire system is considered, however, and when the pros and cons of outside maintenance are weighed, the decision should be for the total maintenance responsibility to be handled internally. Maintenance workers must be adequately trained and tools provided to enable them to perform the required servicing.

Direct Labor

The group most affected by the introduction of CIM is the direct labor or production work force. Their concern is, of course, the possible loss of employment. It is important that the workers who will be displaced are retrained and reassigned. They can be offered an opportunity to upgrade their skills and enhance their eligibility for advancement. The transition of the displaced workers should be made as smoothly as possible.

Indirect Labor

Indirect labor workers, (material handlers and inspectors, for example) are usually less affected than direct labor workers. However, their roles may be modified and it is important that these changes be described to them.

Other Preparations

In addition to the site preparations and communications to prepare people for the system, there are a number of technical, logistical, and administrative preparations that must be made. These include training, engineering documentation, logistical support, and administrative functions.

It is first necessary to identify the personnel who will be assigned the responsibility for maintaining the CIM system and to then assess their credentials. Following this, general and specific training must be provided. The general training may cover basic electronics, computer systems, electronic control, and such things, if the personnel involved do not have adequate knowledge in these areas. Specific training should cover the various elements of the CIM system.

Most CIM manufacturers have developed comprehensive training programs, which they will conduct at either their own facilities or at the user's. Although on-site training creates fewer logistics problems, it is essential that actual hardware be available for hands-on sessions.

It is extremely important that the skilled trade workers be provided with thorough training on the CIM system and are provided manuals and other documentation, special tools, diagnostic equipment, and adequate spare parts.

Engineering Documentation

Manufacturing and industrial engineering personnel have been represented throughout the CIM implementation process as members of the survey team. Now is the time for these organizations to begin reprocessing and redocumenting the operation to which the system will be applied. The documentation normally developed to describe details in the manufacturing operation, the parts and tools involved, the work standards, and engineering specifications should be prepared and distributed. This will, in turn, be used to develop the support activities required.

In addition to process documentation, it is necessary to develop instructions and procedures for the daily start-up and shut-down of the CIM system. This document should provide instructions for routine system and safety equipment checks before start-up and for an orderly shut-down at the end of production to facilitate easy start-up at the beginning of the next production period.

Logistical Support

Material control, material handling, and production schedules may be affected by the CIM system on a continuing basis and will undoubtedly be disrupted during system installation and start-up.

At this time, plans should be developed to support ongoing production while the operation involved is interrupted by the system installation. This may require that the operation be run over capacity in advance of shut-down and the excess parts stockpiled. This, in turn, may require running over capacity in other operations or adjustments of inventories of raw materials or in-process parts. These plans should be based on an accurate system installation schedule, developed by the manufacturing manager. A contingency plan should also be developed in case the installation schedule is not met.

On a continuing basis, the CIM system may be operated during times when personnel are not working. This may, in turn, require adjustment of production schedules and material deliveries and removals. In batch manufacturing, it is advantageous to reduce downtime for changeover between parts by running larger batches and/or by scheduling sequential production runs of similar parts that do not require changeovers. Changes in packaging and delivery of parts may be required by the CIM system and may have an impact on scheduling, storage, and material handling. The development of new production schedules and plans should begin at this time, well in advance of the CIM system installation.

Administrative Functions

The administrative and support staff of the organization that should become involved, at this point, are those people who are responsible for employee relations and training.

The training department should develop schedules for training displaced production workers if required. Of special concern are the training programs for the operation and maintenance of the CIM system.

Maintenance should identify those skilled trade workers who will be responsible for the system and work with the training department to coordinate training schedules. Maintenance is also responsible for special tools, test equipment, and documentation for the system and for ongoing training of replacement personnel and supervisors.

Chapter 8

Operation of CIM

This section discusses the operation of a CIM system and is the final step of the CIM implementation process. (Figures 8–1, Figure 8–2 summarize this step.)

CIM, as noted in this book, is a complex system consisting of many interconnected components of hardware and software, as well as limiting resources such as pallets, fixtures, and tool capacity. Operating a CIM system efficiently can be difficult since any decision to allocate some resources to production of one workpiece necessarily affects the resources available to produce all other workpieces. Furthermore, this interaction can be rather complex and not easy to predict. It is, therefore, important to structure this difficult task in a manner that leads to good decisions. This section will help provide such a structure to aid in CIM operational decision making.

For the design and installation phases of CIM implementation, the importance of involving all levels of the organization has been stressed already. Similarly, you should be aware that once the CIM system enters the operational phase, successful functioning will require ongoing activities at all levels of the organization. The activities required are best understood in terms of the classical three-level view of organizational operation.

The first level consists of long-term decision making, typically done by higher management. This involves establishing policies, production goals, economic goals, and making decisions that have long-term effects. The second level involves medium-term decisions, such as setting the production targets for each part for the next month. These decisions are typically made by the CIM line supervisor, aided by decision-support software. The third level involves short-term decisions, such as which workpiece should be introduced next into the system. Under normal circumstances, these decisions are made by the CIM control computer(s).

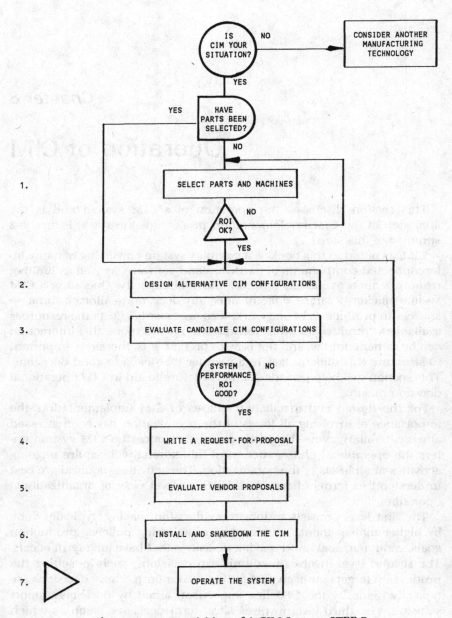

Figure 8-1.   Acquisition of A CIM System: STEP 7

- Schedule parts.
- Batch production if necessary.
- Allocate parts and tools to machines.
- Balance machine loads.
- Use a decision support system to optimize daily operations in the face of machine failure and changing part requirements.

Figure 8-2.   Step 7: Operate the System

However, when an exception occurs, such as a machine failure, the CIM line supervisor may decide to take over some of this decision making, again aided by the decision-support software.

A summary of the three levels of decision making, and associated software, hardware, and management tasks, is given in Figure 8-3. The remainder of this chapter will describe these tasks. The primary aim here is to give an understanding of the issues involved in operating a CIM system and the typical software decision aids that should be available to the CIM managers/supervisors.

A key point is the importance of software decision aids in successful CIM operation. The complexity of the CIM operation task should not be underestimated: Even experienced shop-floor supervisors find that running a CIM system efficiently can be very difficult. In reading the

Time Horizon	Management Level	Typical Tasks	Typical Decision Support Software Used	Hardware Used
Long Term (Months/Years)	Upper Management	•Part-mix changes •System modification/ expansion	•Part Selection Program •Queueing models •Simulation	•Mainframe Computer or DSS computer
Medium Term (Days/Weeks)	CIM Line Supervisor	•Divide production into batches •Maximize machine utilization •Respond to disturbances in production plan/ material availability	•Batching and Balancing Programs •Simulation	•DSS Computer or •CIM Computer
Short Term (Minutes/Hours)	CIM Line Supervisor (exceptions only)	•Work order scheduling and dispatching •Tool management •React to system failures	•Work order dispatching program •Operation and Tool Reallocation Program •Simulation	•CIM Computer

Figure 8-3.   Levels of CIM Decision Making

following sections, take note of the role played by software decision aids so that the development or acquisition of an integrated decision support system (DSS) can be understood to aid in the efficient running of the CIM system.

First-Level Operations

These should encompass the following operational areas:

• Strategic decision making;
• CIM performance;
• Ancillary support.

The execution of activities at this level typically will be supported by software on a mainframe computer. In some organizations, a viable alternative is to have a separate medium-sized computer for these activities, which can be considered a DSS computer.

The activities that need to be performed in each of the operational areas are now described further. In the descriptions that follow, we assume that the CIM system is the major part of manufacturing environment, so the following functions are already being performed at the corporate level:

• Operational MRP System;
• Operational production plan;
• Operational data base management and information system.

These operational functions typically will set overall targets and production goals for a long time horizon. This information usually will reside in the mainframe corporate/plant computer and will serve as inputs to the three levels of operations described in the following sections.

Strategic Decisions

Even after the system is operational, upper management will continue to make CIM-related decisions which have far-reaching consequences. Examples of such decisions are: parts-mix changes, for example, allocating a new product type (or part types) for production on the CIM system; system modification/expansion, for example, adding new manufacturing capability or changing the plant layout. These decisions typically involve complex tradeoffs between economic investments and resulting changes in system performance.

Evaluating CIM Performance

It is important to have a system for monitoring CIM performance relative to management goals and to ascertain the economic (and other) returns from its operation. Real-time CIM monitoring takes place at the third level; here we are concerned with summary indicators of system performance. This is best done by using an MIS that periodically receives detailed performance information from the CIM computer. Analyzing and summarizing this detailed information can then be done according to the wishes of upper management, using the standard capabilities of an MIS.

Ancillary Support

The ancillary support services described here involve all three levels of the hierarchy. However, it is appropriate to consider them as part of level one operations, because they are primarily concerned with developing new applications for the CIM system, and hence have a longer time horizon. Such support services include: extended part programming facilities and part-program verification tools.

Upper level technical support for a CIM system is an ongoing task, because each time a new part type is allocated to production it requires on-going analysis and preparation to be included in the CIM system part repertoire. It is also necessary to refine methods as operational data is accumulated.

Part-program production uses almost every computer in the system. Part design and manufacturing analysis is an effort that must be supported by a variety of utility programs on the mainframe computers. In particular, a part-programming language processor is used to create the part programs from drawings and specifications. To ensure that the programs do what is intended, extensive part-program verification aids must also exist. These usually involve some form of graphic analysis and tool path plotting. In a CIM system where parts may be partially produced on dissimilar machines, separate part programs will have to be used for each machine. Thus, the total effort for part programming can be a lengthy process.

Second-Level Operations

This level encompasses decisions typically made by the CIM line supervisor over a time horizon of several days or weeks. The main tasks to be performed at this level are:

- Dividing overall production targets into batches of parts;
- Within each batch, assigning system resources in a manner that maximizes resource utilization;
- Responding to changes in upper level production plans or material availability.

Batching and Balancing

Since a CIM system creates a larger integrated manufacturing environment, the inputs and outputs of material to the system must match the overall plant MRP and master production plan. These plans specify various availability dates for raw material and various due dates for completed pieces, as well as quantities to be produced. At the same time, in trying to keep to the overall plans, the CIM manager must satisfy many other constraints, such as limited numbers of fixtures and pallets, tool capacity, available machine time, amount of work-in-process, etc. The task of meeting all the production requirements, while using the CIM resources efficiently, is often complex, and is divided into two stages.

The first stage (which is typically done off-line, once a week or once a month) takes the output of the MRP/production plan for the next week (month) and divides the CIM production into batches. This is necessary since tool capacity limitations usually prohibit simultaneous processing of all candidate parts. Each batch should be designed to have a balanced mix of parts, that is, makes even and efficient use of all CIM resources.

The second stage of work order dispatching (usually done on-line by the CIM control computer) uses the targets set by the batching and balancing stage to decide when to introduce the next workpiece into the system and which part type that workpiece should be. This decision is part of the third level of operations. The first stage introduced previously, which involves batching and balancing the workload on the CIM system, will be discussed in more detail below.

Batching Parts on a CIM System

The need for batching in a CIM system can arise for a variety of reasons. Prime among these are the tool capacity constraints that exist for each machine. If the parts to be produced require more tools than will fit on the machines, they will have to be divided into batches, with tool changes between. In addition, batches could be mandated because internal pallet storage is insufficient to handle all parts at once or because part due dates and casting availability dates are widely staggered.

Assuming that there is enough machine capacity to process all desired parts by their due dates, then it must be possible to split the group into a number of smaller batches. However, questions immediately arise as to how many batches must be formed and what parts should comprise each batch. The problem is complicated when trying to process parts efficiently, on schedule, while minimizing in-process inventory and staying within tool capacity constraints.

If the batching is performed solely using manual procedures, it may take a considerable investment of time to produce workable solutions, for example, solutions where tools needed at each machine do not exceed the machine's tool storage capacity. However, this task can be done much more efficiently using automated decision aids.

The main criterion the batching procedure should satisfy is to minimize the total time it takes to process all parts. This translates to the following two issues:

1. Minimize the number of batches required to process all parts. (This minimizes the time associated with batch changeovers.)
2. Maximize the average utilization over all machines. (This minimizes the time required to work through an individual batch.)

The second of these issues highlights the need for balancing the work evenly among the machines. This problem is addressed next.

Balancing the Workload on a CIM System

The need to arrive at balanced allocations of parts and tools to CIM machines arises from economics. It is important that the expensive resources represented by CIM machines not be allowed to stand idle. Workloads must be balanced so that all machines finish their work for each batch more or less together and a new batch can start immediately.

Typical constraints that influence the allocation of parts and tools to machines are tool capacity constraints, tool costs, fixturing limitations, in-process inventory, system workload, and machine failure statistics. As with batching, there is a complex problem to solve, and manual solution is both difficult and very time consuming.

Two main issues should be addressed in the balancing procedure. They are:

1. Minimize the differences in time required for workload assigned to different machines.
2. Be sure *all* work for each batch is in fact assigned to some machine in the system.

The second issue here brings up the possibility of conflict. Due to individual machine tool capacity constraints, it might in fact turn out to be impossible to assign all the work prescribed for a given batch. This would depend on the batching procedure used. For example, the batching procedure used by one software tool is designed to minimize the chance for such an outcome, but it cannot absolutely prevent it. So there has to be a mechanism for iteration: If balancing fails, batching must be tried again, with some modification to its inputs.

Level-Three Operations

This level is concerned with the detailed decision making required for real-time operation of the CIM system. The time horizon here is typically a few minutes or hours, and the decisions involved are:

- Work order scheduling and dispatching (which part to introduce next into the CIM system, and when);
- Movement of workpieces and MHS (which machine to send this workpiece to next, which cart to send to pick up this workpiece, etc.);
- Tool management;
- System monitoring and diagnostics;
- Reacting to disruptions (failures of one or more system components, sudden changes in production requirements).

During normal system operation, most of these decisions are made by software in the CIM computer. However, when an exception occurs, such as failure of a machine, the CIM supervisor will usually take charge of the decision making. If a machine is going to take a long time to repair, he may, for example, decide to reallocate its production to other machines. This involves a complex sequence of tradeoffs between part production rates and machine and tool capacities. Again, the CIM supervisor's task can be simplified considerably by employing various software decision aids. More will be said on these below. These decision aids should typically reside on the CIM computer, to enable rapid implementation of the changed decisions, but in some systems, the architecture could use a separate computer.

Work Order Scheduling and Dispatching

This task controls the flow of workpieces into the system. It takes as input two sets of parameters. The first set of parameters are those the batching and balancing function in the second level decides, which defines the overall allocation of system resources to production of each type of part. The second set of parameters are supplied by the system manager or the CIM control computer, and they specify the current status of the system, such as failed machines, types of pallets/fixtures currently available, raw material available, deviations from desired production rates, etc. The work order dispatching function takes into account all these inputs to decide when to introduce the next workpiece into the system and which part type that workpiece should be.

While this task is usually carried out by the CIM computer, the actual division of decision making between the supervisor and the computer can vary, depending on the supervisor and on the particular CIM system. Typical systems have the CIM computer making the decisions, but they allow the supervisor to override the computer or supply various inputs that can influence those decisions. An overview of the usual "control inputs" available to the supervisor follows.

- Total number of pallets in the system. Generally, increasing the supply of pallets will increase the rate at which parts flow through the system and vice versa. This is because having more pallets increases the probability that a part will always be ready for processing when a machine becomes idle. There is, of course, a point of diminishing returns. In fact, as more and more pallets are added to a system with limited storage capacity, the resulting congestion may actually reduce throughput.
- Total number of each pallet-type. Most CIM systems operate in a "closed system" mode; when a pallet comes out of the system, a part is chosen that can be fixtured on the pallet and the pallet is then sent back in. If more than one part type can use a given pallet type, the part that is most behind in production is usually chosen.
- Part priorities. Some control systems allow the operator to fine-tune part priority above and beyond allocation of resources to it—machines, pallets, etc. The basic mechanism is to alter the processing order so that certain parts waiting to be processed by a particular machine can be processed first. This is in contrast to arrangements that force a first come, first-serve processing.
- Scheduling interval. Some systems allow a time interval to be specified between successive introductions of workpieces of a particular part. In this case, even if the resources required for a workpiece (pallets, fixtures) become available, the system will not introduce that workpiece until the specified interval is over. (Of course, if the interval is over but resources are not available, the introduction will have to be delayed until the resources become available.)

In the case of each of these "control inputs," it is not easy to predict precisely how changing the value of an input for a part will affect production of that and other parts. Here again, software decision aids can prove very helpful to the system supervisor. The most reliable software tool for predicting the consequences of any change in inputs is a detailed simulation program. However, if many different options are to be tested, this can sometimes take a lot of computer time. Alternative software tools which give more rapid results, but are less accurate, are based on the network-of-queues theory.

Movement of Workpieces and Material Handling System

The decisions as to which machine (or machine choices) is (are) available for a particular operation of a part are made in the batching and balancing stage (level two). However, the real-time movement of workpieces around the system is controlled by the CIM computer, in conformance with the decisions made at the batch/balance stage. Under normal circumstances, this workpiece decision making should be transparent to the CIM supervisor. When an exception occurs, the supervisor may intervene in this decision-making level.

Tool Management

This operational area is concerned with three functions:

1. Collecting and updating data regarding the tools on each machine.
2. Keeping track of tool wear, and replacing tools.
3. Reacting to tool breakage.

The first function involves an interface between the tool crib and the CIM computer. Tooling data is generated in the tool crib and entered by the tool setters into the computer through a terminal located in the tool crib. Alternatively, tool gauging data can be sent directly to the machine tool controller by an automatic tool gauging station. This latter feature can reduce tooling cost by reducing the possibility of manual data transcription errors.

A software module in the CIM computer usually can perform the second function directly. When tooling data is entered for a new tool, a conservative estimate should also be included to initialize the limit at which the tool should be replaced/resharpened. If automatic tool changing is not part of the CIM system; a history of tool utilization will be kept and the tool replacement point adjusted either up or down, based on the results of periodic workpiece inspection. When the use of a particular tool has exceeded its anticipated lifetime, a warning will be issued to the system operator and operations will be prevented until the tool is changed or the warning is overridden. Each time a tool is used, a tooling data file is updated so that current reports on tooling status are available.

It should be noted that coordination of the tool replacement task, the activities of the tool crib, and the delivery of tools from the crib to a CIM machine is a complex task. The system supervisor should devote some planning to this task since the successful operation of the CIM system, that is, meeting the production schedules, depends as much on tool management as it does on management of other CIM resources.

System Monitoring and Diagnostics

These functions are essentially performed under CIM computer control and do not require the CIM supervisor. It should be noted, however, that good monitoring and diagnostics are important for the successful operation of an automated system such as a CIM system, since they indicate areas where intervention may be needed. To the extent that diagnostics show the need for corrective action, the CIM supervisor's role is discussed next.

Reacting to Disruptions

Disruptions in system operation are certain to arise and will require the CIM supervisor to take corrective action. Examples of such disruptions would be:

• Machine failure;
• Tool failure;
• Tool replacement warning;
• MHS failure;
• Lack of component parts.

In the case of a tool failing or needing replacement, the action to be taken is clear. This is not so in the case of machine failure.

There are two courses of action possible when a machine fails: either shift the production of affected parts to another machine or temporarily stop their production. If alternate tooling already exists elsewhere in the system, a shift is easily performed and may automatically be handled by the vendor-supplied control software. If the tooling is not available, work can still be transferred from one machine to another, but so must the tools. The problem is that shifting tools from a failed machine will often displace other tools and their associated parts from a working machine. Some of the many questions that must be considered when making a tool change decision are:

- How will the production of other parts be affected?
- How long will it take to change tools?
- How long will the machine be down?
- Is there enough room to store semi-finished parts?

As before, software decision aids can prove very useful in answering these and other questions and in helping the supervisor arrive at a good decision. In this case, the software tools required would be simulation and/or queueing models.

Similarly, in the case of MHS failure, if the repair is going to take considerable time, the supervisor can decide whether to operate a subsection of the CIM system to produce a subset of the parts. This decision would also be enhanced by use of appropriate software decision aids.

A barrage of such disruptions can force the supervisor into a "fire fighting" mode. The software decision aids described, while useful in predicting the effect of any decision, do not automatically find a good decision. Thus, the supervisor may sometimes have to make many attempts before a satisfactory decision is found.

Integration of CIM Operational Levels

The preceding sections described the various levels of decision making relevant to successful and efficient CIM operation. Figure 8–4 summarizes the decisions involved at each level (only the major decisions are shown, for clarity).

Of equal importance as the decision making *within* each level, is the question of communication *between* the levels. Be sure of the answers to the following questions before becoming "locked in" to a particular system architecture:

- How will data (such as part programs) be moved from the mainframe computer to the CIM computer?
- How will information (such as system performance) be communicated from the CIM computer to the mainframe computer?

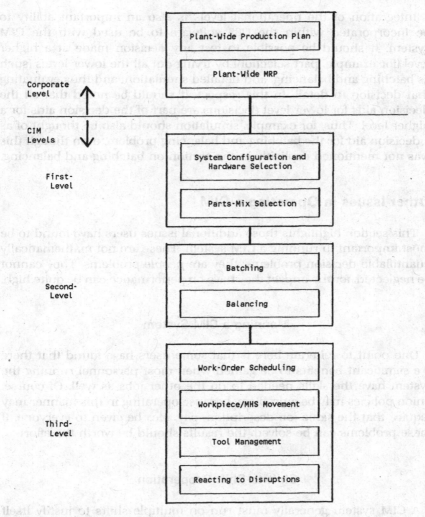

Figure 8-4.   Integration of CIM Operational Levels

• Will a separate DSS computer be used, and if so, how will it communicate with the above two computers?

From the descriptions of the tasks in each level, the importance of software decision aids for enabling a supervisor to run a CIM system efficiently is obvious. If the system is not supplied with an adequate DSS, the creation of one should receive top priority.

Integration of the operational levels is also an important ability to be incorporated within the DSS software to be used with the CIM system. It should be possible to test any decision made at a higher level (for example, part selection) by trying out all the lower levels (such as batching and balancing, and detailed simulation) and thus evaluating that decision in detail. In this respect, it should be noted that all the decision aids for lower level decisions are part of the decision aids for a higher level. Thus, for example, simulation should also be thought of as a decision aid for the batching and balancing problem, even though this was not mentioned explicitly in the section on batching and balancing.

Other Issues in Operating a CIM

This section highlights those additional issues users have found to be most important in running a CIM system. These are not mathematically quantifiable decision problems; they are *people* problems. They cannot be neglected, for the impact they have on performance can be quite high.

Manning a CIM System

One point to reiterate here is that some users have found that there are significant benefits to be gained when most personnel running the system have the skills needed to do the other jobs as well. Of course, union policies may be a problem, because operating in this manner may require that the same job description, pay, etc., be given to everyone. If these problems can be solved, the results should be worth the effort.

Shift-to-Shift Cooperation

A CIM system generally must run on multiple shifts to justify itself economically. This creates a need for the cooperation of personnel between shifts. Users have found this to be an important area and one which is often overlooked.

It is most important to share information such as production attained during the shift, problems encountered, and problems that may occur on the following shift. It is also desirable that similar system operating policies be followed from shift to shift. For example, it is usually detrimental to the achievement of overall production goals if the first- and second-shift operators try to optimize production by continually changing part priorities when the third-shift operator adopts a "hands-off" approach and allows the system to run itself.

Real-Time Part Programming

Eventually, an on the shop floor change to a part program will be needed to prevent a temporary reduction in part production. It may happen, for example, that the third shift finds an extra half inch of stock on its castings. Unless personnel on hand are authorized to make necessary changes, the castings would not be machined.

The risks associated with having a number of people making changes to a program are well known. Success depends upon the cooperation of all involved. Changes have to be approved and recorded.

At this time, there seems to be no clear direction to take for real-time part programming. Users have adopted both approaches, where CIM operators have been allowed to make real-time changes to the part program and where they have not, with varying degrees of success.

Developing and implementing a team approach for all personnel is the only way to successfully use CIM technology.

Part III

Technology Assessment

Chapter 9

Computer-Integrated Manufacturing Technology Assessment

Although all the interrelationships and long-term implications of advanced manufacturing technologies, including computer-integrated manufacturing, are not well understood, the direction of future developments is relatively clear. This chapter describes the technologies that are likely to have a major impact on manufacturing competitiveness, indicates the ways in which those technologies interact, and identifies additional research needs. The technologies have been divided into material handling, material transformation, and data communication and systems integration. These categories are highly interdependent, and divisions are not always distinct, but this categorization provides an effective structure for a broad overview of the major technologies.

Material Handling Technology Trends

This section assesses the major trends in material-handling technology for computer-integrated manufacturing systems. Material handling systems are used to enhance human capabilities in terms of speed of movement, weight lifted, reach distance, speed of thought, sensory abilities of touch, sight, smell, and hearing, and the ability to deal with harsh environments. In this area, it is important to distinguish between equipment technology and design technology. Equipment technology is categorized by its primary functions: transporting, storing, and controlling materials.

221

Transporting

The material handling function of transporting material has been affected significantly by two trends—toward smaller loads and toward asynchronous movement. The former is the result of the drive to lower inventory levels through just in time production. It has been manifested in the development of numerous equipment alternatives that have been downsized for transporting tote boxes and individual items rather than the traditional pallet loads. The inverted power-and-free conveyor, powered by linear induction motors for precise positioning and automatic loading and unloading, is one example of the trend to develop transport equipment for small loads. Automatic guided vehicles (AGVs) for transporting individual tote boxes are also being developed, as are specially designed conveyors and monorails for tote box movement.

The use of asynchronous movement in support of assembly has existed for many years. For example, asynchronous material handling systems were prevalent in automotive assembly before the paced assembly line was adopted at Ford in the early 1900s, and the concept has been applied recently in some European automotive assembly operations. In the early 1970s, Volvo began using AGVs to achieve asynchronous handling in support of job enlargement. Asynchronous material handling equipment is often used to allow a worker to control the pace of the process.

The trend toward asynchronous movement appears to be partially motivated by the apparent success of Japanese electronics firms in using specially designed chain conveyors that place the control of the assembly process in the hands of the assembly operators. A workpiece is mounted on a platform or small pallet which is powered by two constantly moving chains. The platform is freed from the power chain when it reaches an operator's station. After work is completed on the workpiece, the operator connects the pallet to the chain, and it moves to the next station; if the next station has not completed its work on the previous piece, the pallets accumulate on the chain.

Asynchronous alternatives including using AGVs as assembly platforms and for general transport functions; "smart" monorails for transporting parts between workstations; transporter conveyors to control and dispatch work to individual work stations; robots to perform machine loading, case packing, palletizing, assembly, and other material handling tasks; microload automated storage and retrieval machines for material transport, storage, and control functions; cart-on-track equipment to transport material between workstations; and manual carts for low volume material handling activities.

Storing

The major trends in material storage technology are strongly influenced by the reduction in the amount of material to be stored and the use of distributed storage. Rather than installing eight to ten aisles of automated storage and retrieval equipment, firms are now considering one- and two-aisle systems. Rather than being designed to store pallet loads of material, systems are designed to store tote boxes and individual parts. Also, rather than a centralized storage system, a decentralized approach is used to store materials at the point of use.

Among the storage technology alternatives that have emerged are storage carousel conveyors; both horizontal and vertical rotation designs are available. Furthermore, one particular carousel allows each individual storage level to rotate independently, clockwise or counterclockwise. A further enhancement of the carousel conveyor is automatic loading and unloading through the use of robots and special fixtures.

A number of microload automated storage and retrieval systems have been introduced in recent years. The equipment is used to store, move, and control individual tote boxes of material. Rather than performing pick-up and deposit operations at the end of the aisle, the microload machine typically performs such operations along the aisle, since it is used to supply material to workstations along each side of the storage aisle.

Of particular interest has been the introduction of storage equipment for production applications that previously was used for document storage in office environments. The trend toward lighter loads has resulted in a shift of technology from the white-collar environment to the blue-collar environment.

Despite the apparent need for automatic storage and retrieval of individual items, few equipment alternatives are currently available, and those that are have not gained wide acceptance. This particular void in the technology spectrum has existed for a number of years and does not appear to be of current interest to material handling equipment suppliers.

Controlling

The ability to provide real-time control of material has elevated material handling from a mundane "lift that barge, tote that bale" activity to a high-tech activity in many organizations. The control aspect includes both logic control and the physical control of material. In the area of logic control, the ability to track material and perform data input-output tasks rapidly and accurately has had a major impact on material handling. In physical control, automatic controls have been added to a number of material handling equipment alternatives, allowing automatic transfer and assembly.

Perhaps the fastest growing control technology today in material handling is automatic identification. Likewise, the expectation is that the greatest impact in the future will come from the application of artificial intelligence to transporting, storing, and controlling material. Among the alternative sensor technologies available to support automatic identification are a wide range of bar code technologies, optical character recognition, magnetic code readers, radio frequency and surface acoustical wave transponders, machine vision, fiber optics, voice recognition, tactile sensors, and chemical sensors.

The growth in the use of bar code technologies is due to three developments: bar code standardization, on-line printing of bar codes, and standardized labels. The standardization of codes and labels came about through a concerted effort by the user community. The Department of Defense led the way with its LOG-MARS study; that success was followed quickly by a concerted effort by the automotive industry (the Automotive Industry Action Group).

Additional developments in the control of material handling equipment include off-wire guidance of AGVs. AGV technology is one of the most prominent areas of current material handling research and implementation. Functioning as a mobile robot, the AGV is being given enhanced sensor capability to allow it to function in a path-independent mode. Through the use of artifical intelligence techniques, the AGV will be able to perform more than routine transport tasks without human intervention. Using sophisticated diagnostics, it will be able to execute advanced tasks such as automatic loading and unloading of delivery trucks.

A number of European and Japanese firms are making major investments in the development of future generation AGVs. Ranging from vehicles capable of transporting loads in excess of 200,000 pounds to those designed to transport individual printed circuit boards, a number of new entries into the U.S. market are expected within two to three years.

A related control development that will have a major impact on material handling equipment technology is interdevice communications. The manufacturing automation protocol (MAP) being developed by a number of firms led by General Motors is expected to provide the common data transmission link by which many different types of manufacturing hardware, including material handling equipment, will communicate. The driving force behind this standardization effort is the desire for truly integrated manufacturing systems across the entire hierarchy of manufacturing.

Design Technology

In addition to the development of new and improved material handling equipment technology, new thinking has emerged on the design of material handling systems. Specifically, computer-based analysis, including the use of simulation and color graphics-based animation, is being used increasingly to design integrated material handling systems.

Interactive optimization and heuristics also are being applied in the design of material handling systems. Considerable research has been performed in developing performance models of a variety of equipment technologies. Trade-offs between throughput and storage capacity, optimum sequencing of storage and retrievals, and the automatic routing of a vehicle in performing a series of order-picking tasks are some of the issues that have been addressed in an attempt to gain increased understanding of material handling in the future mechatronic factory environment.

Developments in Material Transformation Technologies

This section describes the technologies of individual computer-controlled equipment, from numerically controlled (NC) machine tools and smart robots to computer-aided design and artificial intelligence, developments whose impetus comes from rapid advances in microelectronics and computer science. Rapid developments in very large-scale integration of integrated circuits have reduced the size, cost, and support requirements of information and machine intelligence while greatly increasing its capabilities. Microelectronic technology in the future will be embedded in each machine tool and robot and at every node and juncture of computer and communication networks. The capabilities provided by this embedded intelligence will revolutionize operations on the factory floor.

Machine Tools

Although numerical control was invented and applied some 30 years ago, it continues to change the structure of machine tools in ways that still are not fully appreciated. Computerized numerical control (CNC) has replaced the punched paper tape of the original NC tools. As machines were developed specifically for NC, the traditional lines separating machine types began to blur, and two new classes of machines began to develop.

The first class, called machining centers, generally operates with a stationary workpiece and a rotating tool. Feed of the tool in relation to the work can be handled by additional movement of the tool, the work, or both. The last method is necessary for contoured surfaces and in complex cases may require more than three axes of movement, often five, and perhaps as many as eight. These machines primarily perform drilling, milling, and boring operations, but they also can tap, thread, and, when necessary, mill a surface that simulates work produced by turning.

The second new class of machines developed as a result of NC has rotating work and a stationary tool (except for feed). The machines are called turning centers and resemble lathes. They primarily do internal and external turning, drilling (of holes on the center axis), and threading, but many are equipped with powered stations that permit off-center drilling, tapping, and milling.

Both of these new classes of machines can be equipped with automatic tool changing devices and often have automatic work loading, sensors to check on operating conditions, measuring devices, and other features that enable them to operate for long periods on different workpieces with little or no operator attention. With such versatility, a machining center and a turning center working together can perform all of the basic cutting operations on virtually any part that falls within the operating size limits of the machines.

A number of special cutting and finishing processes supplement the basic processes performed by these machines. These include gear cutting, shearing, punching, thermal cutting, grinding, honing, and lapping. Although NC was not generally applied to these operations as quickly as it was to the basic cutting operations, it is now applied to machines for each of them. (Because shearing and punching are done on presses and usually on sheet or plate material rather than on the heavier workpieces used for cutting, they are usually classified as metal-forming operations. Thermal cutting also falls in this class.)

In addition to the application of NC to traditional metal-cutting operations, several new cutting technologies have become important in many applications. The most widely used of these is electrical discharge machining (EDM), in which the workpiece is precisely eroded or cut by electric pulses jumping between an electrode and the workpiece in the presence of a dielectric fluid. Electrodes, usually made of brass or carbon, are machined to the desired form. Although the cutting process is slow, the machines operate with little or no attention, and EDM is an efficient method of cutting many types of dies.

A major recent development in EDM is the wire cut machine, in which the electrode is replaced by a fine wire sprayed with dielectric fluid. The wire slices through the workpiece as if through cheese, making shaped cuts as the workpiece table moves by NC. The wire is constantly moving between two spools so that, in effect, fresh electrode is always being used.

Low power lasers began to be used for precision measurement about 20 years ago. Higher power lasers are now used for welding and for sheet and plate metal cutting. Within the past two years, precision machine tools that use the laser as a cutting tool have been introduced in the United States and Japan, both for drilling and for cutting contoured surfaces. Other new technologies include the use of electron beams for drilling and welding and the use of plasma flame for cutting.

Parts can be formed from sheet or plate in a variety of presses that bend, fold, draw, punch, and trim. The average age of presses currently in use is much higher than the average age of cutting machines, and users have generally been slower to innovate, but some press-working shops have taken advantage of new technologies. For example, some shops have installed lines in which coiled sheet is unrolled, flattened, trimmed, and shaped into parts by stamping, bending, and drawing in a continuous series of operations. Others have installed transfer presses which make finished parts from strip metal in a continuous series of operations. Much of the progress in forming has come through better control of the material to be formed.

The only extensive use of NC in presses has been in punch presses that combine tool changing ability with two-axis positioning of the work for punching, nibbling, trimming, and cutting with lasers or plasma flame. However, NC controls are now beginning to appear on some other types of presses.

Tooling

Cutting tools are made from a variety of materials: high speed steel; carbides of tungsten, titanium, and boron; oxides of aluminum and silicon (ceramics); cubic boron nitride; and synthetic and natural diamonds. Major advances in cutting-tool materials sometimes cannot be fully utilized until machines designed to take advantage of their properties are generally available.

Great progress in cutting tools has been made by applying a coating of one material (in some cases, two or three coatings of different materials) onto a base material. The proliferation of tool materials and coatings has become so complicated that computer software has been developed to aid the process. The resulting tools last longer, stay sharper, and can be used to cut hard materials such as heat-treated steel and abrasive materials such as fiberglass.

As combinations of materials and coatings produce a growing list of tooling options, the variety and volume of tooling requirements can be expected to proliferate. New product designs and performance requirements, product and process specifications, and changing lot sizes will create an ever-increasing need to match specific tooling with specific production applications. To achieve the high quality, close-tolerance production demanded in the marketplace, manufacturers will require a large inventory of tooling to ensure that the optimal tooling is available for all production requirements. Combined with the increased expense of tooling made with rare materials and precision coatings, the costs of meeting tooling requirements will become major factors in capital budgeting decisions.

Improvements also have been made in die and mold materials. More important, however, is the change taking place in the way dies and molds are produced. Traditionally, they have been made by experienced craftsmen with a great deal of time-consuming cut and try in the finishing stages. The combination of newer EDM machines and computerized programming of die-sinking machines is removing much of the cut and try from this process.

Jigs, which serve to position the tool more accurately in relation to the work for drilling or boring, can usually be eliminated when NC machines are used. In fact, one of the major early advantages of NC was the elimination of the production and storage of jigs. Of course, if the jig also serves as a fixture to hold the work on the machine, that function is not eliminated on an NC machine.

Fixturing

Fixtures hold and locate the part being worked during machining and assembly operations. The main considerations in fixture design are positioning the part in the fixture, securing the part while the machining operation takes place, positioning the fixture relative to the machine tool, positioning the cutting tool relative to the part, and minimizing set-up times. New fixturing techniques add flexibility and programmability to minimize set-up time, maximize the flexibility of the machine, and reduce storage requirements for fixtures.

The characteristics of the fixture depend on the process being performed, the shape of the part, and the tolerances required. For example, the workpiece may be subjected to strong vibrations or torque forces during some operations such as milling, while the forces in assembly operations will be much smaller. The fixtures required for these two operations are quite different and virtually incompatible. When a variety of tasks are performed, a large number of fixtures must be developed, stored, and accessed which is a very expensive undertaking.

The need for a large number of fixtures remains a problem even for computer-integrated manufacturing systems that can quickly and efficiently machine a number of different parts within the same part family. The computer-integrated manufacturing can help reduce economic lot sizes and reduce the expense of keeping parts in inventory. Unfortunately, this advantage is restricted by the need to have different fixtures for different parts. The cost of multiple fixtures can account for 10 to 20 percent of the total cost of the system, and the fixtures can sometimes cost more than the rest of the system. Clearly, the full advantages of a computer-integrated manufacturing cannot be realized without the development of flexible fixturing that can conform to different part types and machining operations.

A number of major research efforts are focused on the problem of flexible fixturing, and several solutions have been proposed. One approach would be to automate the current fixturing process, which uses blocks and clamps to align parts accurately. Instead of skilled toolmakers, robots could be used to assemble fixtures on coordinate measuring machines (CMMs). The fixtures would be mounted on standard pallets, permitting robots to load and unload parts easily and allowing easy alignment with machine tools. The CMM could cost $200,000 and vision-equipped robots at least $100,000; the hardware for the fixtures themselves and the software needed to control the robots would add to these amounts. Although the present cost may be prohibitive, this approach would ensure accurate location of the workpiece and it could be used for both machining and assembly operations.

Another approach partially encapsulates the workpiece in a low melting point alloy prior to machining. Encapsulation has been developed specifically for milling gas turbine and compressor blades of irregular shape. The unmachined blade is precisely positioned in the encapsulation machine. Rapidly injected molten alloy surrounds the blade and provides the clamping face, protecting the blade itself. After machining, the alloy capsule is mechanically cracked open. The problem with this approach is that the blade must be positioned accurately in the very expensive encapsulation machine, which requires a different die for each workpiece. This limits flexibility and adds expense.

A third approach is programmable conformable clamps. Developed at Carnegie-Mellon University for machining turbine blades, the clamps consist of octagonal frames hinged to accept a blade. The lower half of the clamp uses plungers, activated by air pressure, that conform to the contours of the blade. A high strength belt is folded over the top of the blade, pressing it against the plungers, which are mechanically locked in place. Accurate alignment can be done manually or automatically with sensors. Although the clamps are limited in the types and sizes of parts they can hold and their large number of moving parts may reduce reliability, they are automatic and ensure accurate alignment.

Another approach is the fluidized-bed vise, in which small spheres are held in a container with a porous floor through which a controlled air stream passes. The spheres behave like a fluid, conforming to the contours of even irregularly shaped parts; when the air flow is stopped, the spheres come together to form a solid mass that secures the part. The advantages of this approach are that a variety of part shapes can be clamped, the clamping process is automatic, and the vise is inexpensive to build and operate. However, additional research is needed to establish a predictive model for the device and to eliminate the need for an auxiliary device to determine the location and orientation of the workpiece in the vise. Research is also under way in which electrically active or thermally active polymers are used in an authentic phase change bed instead of the pseudo phase change of the air-sphere approach.

None of these approaches offers the flexibility needed in terms of variety of applications, the types and sizes of parts that can be held, and expense. They also do not address the problem of locating the workpiece. The first three approaches use mechanical stops or surfaces, and the fourth requires an additional measuring system; this problem may be overcome by combining flexible fixturing devices with sophisticated robots.

Smart Sensors

As machine tool automation advances, the instrumentation on the machine becomes increasingly important. Most of the early problems with automation tended to be instrumentation problems. Sensors to determine what is happening and monitoring systems to evaluate the sensor information are both needed. The role of sensors in a manufacturing environment is to gather data for adaptive control systems, for example, to supply guidance information to robots or to provide measurements for quality assurance and inspection systems. Sensors can provide automated equipment with vision, touch, and other senses, enabling the equipment to explore and analyze its surroundings and, therefore, behave more intelligently.

Sensors are currently used in factories to provide different types of data, such as the bipolar on-off of a limit switch, the simple numeric data of a temperature sensor, and the complex data provided by a vision sensor. Vision sensors, for example, can be used to determine part identification, orientation, and measurement data. Other sensors, such as tactile, acoustic, and laser range-finding sensors, are being used to measure force and shape, provide range data, and analyze the quality of welding processes.

Sensor technology is a very active field of research. Sensor research that shows promise for manufacturing includes micromechanics, three-dimensional vision for depth sensing, artificial skin for heat and touch sensing, and a variety of special-purpose sensing devices. Some of the special-purpose devices have no human analog. Examples are the water vapor sensors being developed for use in sophisticated adaptive-control algorithms and the optical laser spectrometry probes that monitor chemical processes in real time. The use of adaptive closed-loop control systems in manufacturing has increased the demand for a wide variety of special-purpose sensors and has stimulated the demand for sophistication in general-purpose sensors such as vision sensors.

Other research is focused on the analysis, interpretation, and use of the data provided by sensors. Through the use of Very Large Scale Integrated circuits (VLSI) techniques in Integrated Circuit (IC) fabrication, intelligent sensors equipped with microchips can process data even before it leaves the sensor. For example, research is under way on machine vision systems that can inspect IC wafer reticles. Research on this machine vision system is focused on the mechanical accuracy of positioning devices, on the interface to the CAD data base describing the reticle, and on modeling the fabrication process to predict what the machine vision system will see. The visual information itself must be interpreted to determine whether to accept the wafer under inspection or to identify the flaw and provide feedback to correct for any imbalance. This type of smart sensor will eventually be integrated into many elements of manufacturing.

Model-based sensor systems such as these which use process, CAD, simulation, and control algorithms are expected to provide manufacturing sensor systems of the future with very complex analysis capabilities. These analysis capabilities will far surpass the monitoring and control capabilities of human operators by being more sensitive, more precise in analysis, more rapid in feedback response, and more precise in corrective action. They will allow the mechatronic factory of the future to work to very fine tolerances while maintaining consistently high quality control, approaching zero defects.

Smart Robots

One of the most common uses of advanced sensors is to make robots smarter. The sensory control of a robot are the sensors in its work cell that provide information to the robot's central controller. The "intelligence" of the robot is determined by the combined capabilities of its controller, its sensors, and its software. Most of the robots in the world's factories today have primitive controllers and software and few sensors. They mindlessly weld, paint, and pick-and-place, and some would continue to do so even if no object were present to paint, weld, or grasp. Such robots are locked into a predetermined program that does not adapt to unexpected changes in the work cell. In contrast, advanced robot systems have smart sensors that inform the robot of the state of its world, controllers that can interface with the advanced sensors, and software that can adapt the robot's program to reflect the changing state of its world. This is an example of adaptive behavior using a closed-loop feedback system; to a degree, it is what people do when they engage in behavior that uses the senses. It is expected that 60 percent of all robots, especially those used for inspection, assembly, and welding, will utilize machine vision, tactile, and other sensors within the next ten years.

Smart robots have many advantages. About one-third of the cost of a robot work cell is the fixturing that holds or feeds each part in precisely the same way each time. This cost can be saved by smart robots that can find the part they need even if it is askew, upside down, or in a bin with other parts; it is easier to change a robot program than to change the fixturing. Smart robots will be much more adaptable to product changes because they will have less fixturing to change. Smart robots will be even more adaptable to different tasks when they can easily change their end effector for a drill, deburrer, laser, or whatever tool is required.

State-of-the-art robot systems embody elements of adaptive control and are now coming into use in factories around the world. One example is arc welding robots whose welding path is planned with the aid of a machine vision system that determines the location and the width of the gap to be welded. The robot software then adjusts the path and speed of the welding tool as the welding progresses. Although the welding example shows how adaptive control enables a robot to perform a task with built-in variance, the variance found in arc welding can be foreseen easily and taken into account by a human engineer or programmer. Adaptive control for robots with less-structured tasks is still in the research stage.

Robots are programmed through a special-purpose computer language. State-of-the-art languages allow the robot to perform limited decision making on its own from information obtained with its sensors. However, these programming languages are limited because they can neither interpret complex sensory data, as from a machine vision or tactile sensor, nor access CAD data bases to get the information they may need to identify the parts that they sense. Present languages are also robot dependent; that is, they do not allow the transfer of programs from one robot to another. This means that robots must be programmed individually by valuable, highly trained programmers.

New robot programming languages that address some of these limitations are in development in academic and commercial research laboratories. The new task level languages will allow robot programming at higher levels: the robot can be told what to accomplish or what to do with the part, and it will determine the best way to accomplish the task. The benefits expected when the new languages reach the factory floor include reducing the cost of programming, facilitating the coordination of two or more robots working cooperatively, and enabling advanced sensors to interface with the new systems.

Computer-Aided Design

Computer-aided design is not a new technology; it has already achieved wide acceptance and use in manufacturing design, and it has replaced traditional drafting techniques in other areas such as architecture and cartography. It is important to understand CAD as a technology because it interrelates with many of the other technologies described here. For example, CAD-type systems are now being used to program robots and NC machining centers.

A CAD system is composed of a graphics terminal on which can be displayed a picture of the part being designed. Designers enter the part data by drawing on a graphics tablet connected to the computer. A keyboard is used to enter dimensions and other data. The part description is then stored as one of many such part descriptions in a CAD data base. The computerized part description is not a picture, but rather a representation of coordinate points and geometric shapes from which a picture can be constructed. The Initial Graphics Exchange Standard has been developed for transferring data from previously incompatible representations on one CAD system to another CAD system.

CAD offers many immediate benefits: parts can be rotated, scaled, and combined onscreen in three dimensions to enable designers to better visualize them; repetitive sections can be redrawn automatically; overlays can be easily shown onscreen; and engineering drawings can be easily updated and printed.

Other, more significant, benefits over the long run concern the use of the data in the CAD data base. These data—a computerized representation of the parts—can be used by the engineering and process planning functions, saving much reentering of data, eliminating sources of human error, and opening up a great avenue for cooperative design that includes feedback from engineering and manufacturing. Also, if the CAD data base is the only and therefore up-to-date source of part specifications, it eliminates a major current problem, concurrent use of multiple versions of part specifications.

The microelectronics industry probably has the most integrated use of CAD. A new microchip can be designed on a CAD terminal. Once the design is in the CAD data base, the chip's performance can be simulated, the design can be modified if necessary, and the masks for the chip can be made, all automatically from the data entered at the CAD terminal. Although other industries have not yet achieved this level of integration, it has become an embodiment of the computerized design-test-modify-test-fabricate model that may change the way manufacturing entities are organized.

System Simulation

As computer-integrated manufacturing systems come to include advanced systems such as smart robots, it becomes more and more important to be able to simulate their behavior. Two distinct kinds of simulation are now being used in manufacturing. One is the simulation, often graphic, of a single process, robot, or work cell. The second is the simulation, generally mathematical, of a system such as a computer-integrated manufacturing, a new or modified production line, or an entire factory. The former may be regarded as important in tactical or local decisions, the latter in strategic or system decisions. For this reason, system simulation will be treated here, and mathematical modeling will be covered later in the communications and systems section.

Robot system simulation is beginning to be used to select the most appropriate robot for a particular task or work cell and to plan the cell layout. The production engineer can use simulation to reject robots that do not visually appear to suit the task because of their arm configuration or timing constraints. System simulation is also used for visual collision detection in the work cell, but this method is still prone to error.

Some vendors of robot simulators have adapted their software to generate actual robot control programs, which is termed off-line programming. It permits the development of robot programs without shutting down a productive work cell, thus allowing efficient, concurrent work cell design. Although programs developed off-line currently must be used on the specific robot for which the system was designed, research to include a variety of robots in the simulation system is being conducted. For example, an off-line robot programming system has been developed that can simulate any of six commonly used robots. Researchers are also working on the related problems of how to simulate a complex sensor, such as a machine vision sensor, and how to debug an off-line robot program that makes decisions based on advanced sensory input. System simulation and off-line programming promise to provide cost, time, and personnel savings in the efficient design of programs, work cells, and processes.

Artificial Intelligence

Artificial intelligence is a set of advanced computer software programs applicable to classes of nondeterministic problems such as natural language understanding, image understanding, expert systems, knowledge acquisition and representation, heuristic search, deductive reasoning, and planning.

Artificial intelligence (AI) technology will emerge as an integral part of nearly every area of manufacturing automation and decision making. Research that will affect manufacturing is being conducted in several areas of AI, including robotics, pattern recognition, deduction and problem solving, speech recognition and output, and semantic information processing. As with simulation, AI will be used at different levels in the factory of the future. Most of the AI applications will be integrated into the software that controls automated machinery, record keeping, and decision making.

Artificial intelligence is not a new field, but the maturing fruits of 20 years of AI research are just now becoming available for commercial applications. The types of AI products that will have a significant impact on manufacturing include:

- expert systems in which the decision rules of human experts are captured and made available for automated decision making;
- planning, testing, and diagnostic systems; and
- ambiguity resolvers, which attempt to interpret complex, incomplete, or conflicting data.

The AI applications that deal with individual machines, processes, or work cells are described here; those that deal with system-level decision making will be integrated into the factory communications and systems technologies section of this chapter.

Expert systems are in productive use today in isolated industries; petrochemical companies, for example, use expert systems for the analysis of drilling samples. Digital Equipment Corporation has used an expert system for a number of years, saving several million dollars annually in configuring the company's VAX computer systems. As human experts with years of experience become scarce, the expert system provides a way in which to capture and "clone" the human expert. An interesting feature of expert systems is that they can explain the train of reasoning that led them to each conclusion. In this way, the systems also can be used to augment human decision making, in much the same way as medical expert systems have been used. Current expert systems are best suited to situations that are somewhat deterministic when the expert's rules are known. For this reason, rapid emergence of expert systems can be expected in limited areas of technical knowledge such as chip design, arc welding, painting, machining, and surface finishing. In the 1990s, expert systems are expected that will learn from experience; this means that expert systems eventually will be developed for specialities in which there are no human experts.

Although still primarily in the laboratory, one type of AI software is attempting to simplify the use and expand the applicability of programmable equipment. For example, advanced user interfaces are now being developed that use "natural language," so that a manager can type a request at his work station in more or less plain English. The AI software will determine what he means, even if the request has been phrased conversationally or colloquially, and provide interactive assistance for decision making. By the year 2000, managers will probably be communicating with their workstations by voice, another application of AI techniques. Artificial intelligence technology promises to make it much easier for computers and computerized equipment to be used by personnel not having computer training, such as managers, engineers, and operators on the factory floor.

Factory Communications and CIM System Technologies

In contrast to the materials handling and process technologies described above, communications and systems technologies tend to operate at higher levels, allowing previously separate areas of manufacturing to be integrated into systems of manufacturing. A computer-integrated manufacturing system is a system created by the interconnection and integration of processes of manufacturing with other processes or systems. This definition implies that computer-integrated manufacturing systems vary from a basic system, which couples a few processes, to a hierarchical system, (Figure 9-1), that integrates lower level manufacturing subsystems into the single aggregated systems. Such a system is defined as a computer-integrated manufacturing (CIM) system, shown in Figure 9-2.

This variation in complexity and level makes the concept of a computer-integrated manufacturing system elusive to grasp. It may be helpful to think of it as an approach, a systems approach, to incrementally integrating the functions of the manufacturing corporation.

The major characteristic of computer-integrated manufacturing systems is their sharing of information, their communication. Traditionally, manufacturing information has been created and communicated by humans writing on paper. This paper information was based on the understanding of the human expert at that moment, although often that understanding did not accurately reflect the real state of the factory at that moment. This paper method is people intensive, time consuming to create and distribute, often inaccurate, and in frequent need of revision. As an information communication method, it virtually guarantees delay, inaccuracy, and expense.

The advent of computer technology and network communications is changing the face of the factory floor, much as office automation has changed the front office. This technology permits the system to generate

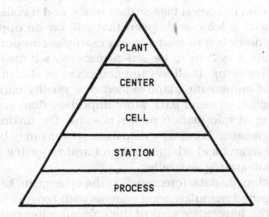

Figure 9-1. Computer-Integrated Manufacturing Hierarchy
Source: Allen Bradley Corp.

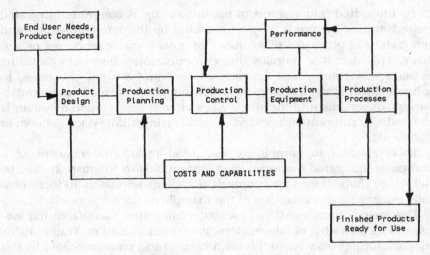

Figure 9-2. Integration functions in a computer-integrated manufacturing system.

its own data according to the information provided by real-time sensors built into automated machining, assembly, and inspection stations. The system gives the data to a computer, which interprets the data and takes appropriate action. This action may be to control the machining process, to replace a worn tool, or to decide whether to communicate the data, to whom, and how much.

This automated creation and sharing of information avoids the present duplication of data in several files or data bases, and it collects and communicates data at a scale and speed that will create opportunities in manufacturing never before available. For example, computer-controlled feedback permits a system to be self-diagnosing, self-maintaining, and eventually self-repairing. It allows the collection of statistical data that can be used for immediate adaptive feedback, quality control analysis, and the production of trend data. More important than any single benefit, this sharing of information makes possible the linking of systems into system aggregates. Previously disparate systems may be linked horizontally, and hierarchical adaptive control and reporting systems may be created by integrating vertically.

This information, or data, integration is the synergistic key to building computer-integrated manufacturing systems with broader scopes and at higher levels. The long-range goal of the computer-integrated manufacturing systems approach is the complete integration of the manufacturing subsystems that operate on the factory floor, the tie-in of techniques of optimization, mathematical modeling, scheduling, and data communication with the other functions (accounting, marketing, etc.) in the total manufacturing enterprise. Note that manufacturing systems are at once a means and an end.

Systems of manufacturing are integrated through the application of several technologies: communication networks, interface development, data integration, hierarchical and adaptive closed-loop control, group technology and structured analysis and design systems, factory management and control systems, modeling and optimization techniques, and flexible manufacturing systems as shown in Figure 9–3. Artificial intelligence techniques will be embedded in, and inseparable from, most of these technologies. Communications technologies—those associated with networks, interfaces, and data bases—may be the most critical to U.S. manufacturing progress because they are the keys to the immediate development of manufacturing systems. On the other hand, technologies that analyze, manage, and optimize the system hold the greatest promise for improving the long-term competitiveness of U.S. manufac-

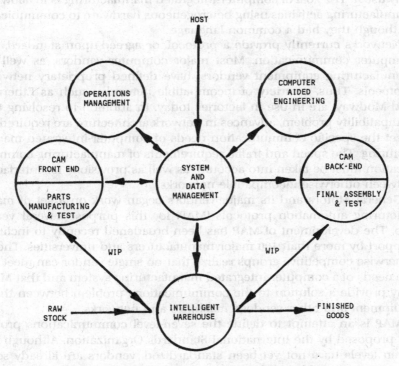

Figure 9-3. Interrelationships of CIM Elements

turing. These technologies will facilitate progress toward the goal of total integration from design to delivery. Each will be described in depth, both individually and as they relate to the full computer-integrated manufacturing concept.

Networks

The manufacturing network will be the backbone of factory communications and, therefore, of factory automation. Communications between tightly coupled components, such as robots and sensors, and between elements of a computer-integrated manufacturing require that data be exchanged in real time. As the complexity of the factory system increases, including the linkage from design to planning and production, the need for factory communications will continue to expand. Networks provide the physical mechanism for this communication between heterogeneous systems. The network must not only transmit the raw data but also retain its meaning, so that a different computer, running a different program, may use it. The goal of computer-integrated manufacturing is to allow all manufacturing activities using heterogeneous hardware to communicate as though they had a common language.

Networks currently provide a protocol, or agreed-upon standard, for computer communication. Most major computer vendors, as well as manufacturing equipment vendors, have defined proprietary network protocols. Thus, a variety of incompatible networks, such as Ethernet and Modway, are in use in factories today. In addition to resolving the compatibility problem, advances in network architecture are required to meet the specific communication needs of computer-integrated manufacturing. The speed and traffic requirements of manufacturing communication must be taken into account, as well as provisions for interfaces between otherwise incompatible networks.

General Motors and its major vendors began work on a set of manufacturing automation protocols (MAP) for this purpose several years ago. The development of MAP has been broadened recently to include support by more than 100 major manufacturers and universities. These otherwise competitive groups realize that no single vendor can meet all the needs of a computer-integrated manufacturing system and that MAP may provide a solution to the communications problem between their equipment and other vendors' machines and networks.

MAP is an attempt to define the seven-level communications protocol proposed by the International Standards Organization. Although all seven levels have not yet been standardized, vendors are already selling MAP-compatible products, and farsighted purchasers are demanding that their new hardware be MAP compatible. A recent breakthrough by Industrial Networking, Inc., has put MAP on a single microchip, which will facilitate the development of factory communication networks among heterogeneous machines. While significant challenges remain, the broad membership and participation in the MAP effort can be used as a model for specifying and solving other manufacturing system problems.

Interface Standards

The network is expected to provide the physical and logical path for data communication in a factory system, but much more is required for effective communication. Networks provide the physical language and format, but do not address the semantics or effective use of the information communicated. Interface standards are needed to facilitate the effective communication of meaningful data.

The key to data integration is standardization that does not stifle innovation. Standardization of data representation within the data base is necessary to allow the full meaning of the data to be retained even when it is communicated. Current practice requires vendors of systems or modules to provide special-purpose interface definitions for each pair or family of modules that communicate. However, in some areas standards have evolved through the cooperation of users and vendors. Examples are the CLDATA file for NC machines and the Initial Graphics Exchange Standard (IGES) for CAD data base information exchange. IGES has enabled previously incompatible CAD systems, with data stored in radically different formats, to communicate that data while preserving most of the meaning. Yet these standards rapidly grow out of date as technology moves forward. CLDATA is inadequate for nondeterministic (sensor-based) machine tool programs, and IGES does not work on solid modeling CAD systems. The IGES continues to evolve, pointing the way to wider data integration. The challenge is to define standards that will withstand the demands of continued factory innovation or to establish mechanisms to update standards as needed.

Standards are also the solution to the interface compatibility problem that arises when equipment from different vendors is used in a network. The interface connects one machine to a communications system, which is connected to other machines, computers, and communications systems. The RS–232 interface standard is a simple protocol that has allowed communication between heterogeneous microcomputers and between computers and a host of other devices. Many machines already come with the limited RS–232 interface, but more progress is needed in standardizing manufacturing interfaces. MAP includes the definition of intelligent interfaces which can connect previously incompatible systems to a network.

The lack of interface standards can be a major impediment to achieving computer-integrated manufacturing. If well-defined information interfaces between modules or subsystems were established for the components of manufacturing systems, components could be developed independently and enlarged as advances in technology became available. This would facilitate compatibility of the equipment of multiple vendors in the heterogeneous systems expected in the factory of the future.

In addition to these interface standards for information, two other kinds of interface standards are needed. The first and most neglected is the interface between human and programmable systems. The second is the physical interface between mechanical systems.

The man-machine interface includes the commands to be given by the human to make the machine perform a task successfully and the input device or physical method—keyboard, teach pendant, joystick, light pen, or voice—for entering those commands. Most current programmable systems are commanded through a programming language that is proprietary to the vendor of the system. This has given rise to a Tower of Babel of control languages requiring highly trained programmers to control modern computer-integrated manufacturing systems. No programmer can begin to master all of the languages and input devices found in a computer-integrated manufacturing factory.

Two recent trends are expected to ease the interface problem between nonprogrammers, such as engineers and technicians, and the increasingly complex programmable automation systems found in the areas of robotics, NC tools, material handling, and processing systems. Hierarchies of languages and personnel are being developed in which highly trained programmers will deal with the raw control languages and sophisticated control algorithms, less-skilled programmers will deal with a higher level simplified language, and equipment operators will not use actual programming languages at all, as shown in Figure 9–4. This hierarchy—automation systems programmer, applications programmer, user programmer, and user—parallels the evolution of personnel in the computer field.

The second trend is the use of AI to develop task-level control languages. Task programming systems will reduce programming requirements to the steps on a common process planning sheet so that programmable manufacturing systems of the future will be controlled by statements similar to those one would give to a person doing the same task. These advances will provide new generations of specialized, user-friendly

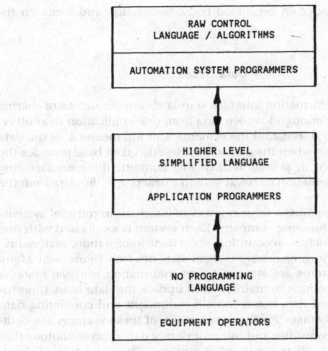

Figure 9-4. Robot User Language Structure

computer-integrated manufacturing subsystems that will make the most of factory personnel.

The mechanical interface problem for the factory is solved most easily by the development and adoption of standards. The lack of standards for the newer systems is a major impediment to progress. Examples of mechanical interfaces in need of standardization include:

- robot end-of-arm and gripper attachments;
- pallets, totes, and other part conveyances;
- the mechanical interface for the loading and unloading of parts and pallets at machining centers; and
- the interface between robot carts and material handling systems.

Progress with the mechanical interface problem requires the usual consortium to agree on and promulgate standards. The major roadblock has been the lack of an organized body, leadership, and focus on the problem.

Data Bases

Network and information interface standards are the means of sharing data, but it is not enough to move data from one application to another. The data must be stored, and the semantics, or full meaning, of the data must remain intact when the data are retrieved. A data base provides the long-term memory, or storage facility, that contains the manufacturing data, and the information retrieval system extracts specific data from the data base.

Current practice finds a large number of information retrieval systems in place even at the same company. Each system is associated with one major function, such as accounting, shop floor information, material requirements planning, or quality control statistics (see Figure 9–5). Many of these data systems are state-of-the-art information retrieval systems with complex functions to maintain and update the data base. Unfortunately, the different data bases contain redundant and conflicting data in incompatible formats. Furthermore, many of these systems are dedicated to a single computer and use proprietary data representations that are incompatible with those of other systems. Thus, the task of developing an integrated manufacturing data base management system that can include every major function in a factory is formidable.

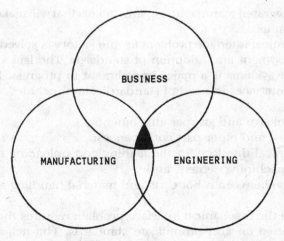

Figure 9-5.   Interrelationship between information data bases

The most serious immediate barrier to the integration of manufacturing data is the incompatibility of CAD data with information needed by computer-aided engineering (CAE) and process planning. CAD technology has become highly effective in capturing the geometry of parts, including the description of dimensions, shapes, and surfaces. Real parts, however, are made up of smaller components and may themselves be components of a larger assembly. The CAD data base currently cannot capture the relationship of the parts to the whole, but both CAE and process planning require detailed attention to the joining of separate parts, their mating surfaces and tolerances, and their overall dimensions after assembly. The CAD data base does not include knowledge or specification of materials, but CAE needs material data for its engineering analyses, process planning needs it in the creation of NC programs, and material handling needs it to select material from inventory.

A second serious barrier to the integration of manufacturing data is the current inability to model the processing portion of the overall system. Such a model would allow information on production costs and capabilities to be fed back, on-line, to the product design activity as it is being performed. This capability is essential to optimizing the producibility of products at the design stage. With this capability, each decision proposed in the engineering design process would result in simultaneous information on the effect of that decision on production costs and required capabilities (relative to available capabilities) for production of the product. It also would result in major cost savings in the production activity, since it is well known that the majority of production costs are frozen at the engineering design stage. While such computer-based integration of manufacturing data is technologically feasible, many difficult problems must be solved to bring it into being.

A further problem with the data in a CAD data base is that the geometrical data cannot be searched or aggregated in the ways that have become standard for textual data. Without explicit hand coding, it is not currently possible to retrieve all parts that use a particular fastener. Group technology is an attempt to code and classify the geometry, function, and process data in a way that will permit the use of standard retrieval functions.

One of the keys to the data integration problem lies in the development of flexible data schemas. A schema is a method of storing data so that its meaning and accessibility are retained. Most data bases use rather fixed data schemas that restrict the new types of information that may be added and limit data retrieval capabilities. Future data bases will have more flexible schemas so that, for example, materials information can be added to the CAD data base by an engineer at a CAE station or by an expert system that contains knowledge of materials and applications.

Beyond the compatibility problem are other technical challenges to the implementation of computer-integrated manufacturing data base systems. For example, experts predict that future computer-integrated manufacturing data bases will be 20 to 50 times larger than present data bases. The size of the data base, the time tolerances for communication, and the variety of users suggest that a manufacturing system data base will be distributed across multiple heterogeneous systems, which may be in different geographical locations. This presents significant technical challenges to the achievement of a logically integrated manufacturing data base. The concepts and protocols normally used to ensure proper access, control, and update will need to be expanded to meet this sophisticated method of data base organization. Interim solutions in place today are neither geographically nor heterogeneously distributed, but progress toward these goals is being made.

A last challenge posed by the manufacturing data base is the use of probabilistic or incomplete data. Current data base systems can only represent facts and cannot deal with uncertainty or conflict within their data. Manufacturing information systems of the future will depend on AI to deal intelligently with this type of information.

One of the most critical roles of people in the factory of the future will be to interact with intelligent manufacturing systems through workstations, terminals, or networked microcomputers. As expert systems and other forms of AI become embedded in systems of manufacturing, the systems will be able to perform more and more of the decision-making tasks previously performed by people. At first, these automated decisions often will have to be reviewed by people and then interactively modified, much as an architectural plan takes shape in a dialogue between client and architect. People without knowledge of the data schema will routinely query the system for information needed to make decisions. The data retrieval system will have to determine exactly what is important to the inquirer and then retrieve and massage the appropriate data. The person may even want the system's "opinion," or the system may ask for the person's opinion. The mechatronic factory of the future will regard personnel and intelligent systems as partners in a dialogue that should encourage very sound decision making.

Group Technology

Group technology (GT) is a key philosophy in the planning and development of integrated systems. In practice, GT is defined as a disciplined approach to identifying by their attributes things such as parts, processes, equipment, tools, people, and customer needs. These attributes are then analyzed to identify similarities between and among things; the things are grouped into families according to similarities; and these similarities are used to increase the efficiency and effectiveness of managing the computer-integrated manufacturing process.

Although it is relatively simple to define GT, it is difficult to create and install a GT system because of the difficulty in defining clearly how similar one part is to another. For example, parts can be categorized in terms of shape or manufacturing process requirements. These two different viewpoints require a flexible approach to the GT data base and the realization that parochial, departmental views of coding may allow some localized cost saving but miss the large corporate savings possible.

The GT concept requires that the attributes of a thing, such as a part, be identified and classified. Attributes can be visual, such as the surface finish or shape of a part; mechanical, such as the strength of the material; or functional, such as the clock aspect of a printed circuit board. The attributes may also be related to the environment of the part, such as the processes or equipment necessary to make it. Because of the many possible coding strategies, it is hard to know in advance exactly what attributes will be important as the GT data base is used by more and more kinds of software. It is therefore important to guarantee that the data base structure is flexible enough to add attributes and to modify coding schemas as necessary for new applications.

The four basic GT applications are design retrieval, computer-integrated manufacturing, purchasing support, and service depot streamlining. Design retrieval is a GT application in the design engineering area to provide the maximum potential for part standardization. Moreover, it permits greater cooperation between the design engineer and manufacturing engineer by providing feedback about specific part attributes in the GT data base. Design retrieval will also supplement product reliability data based on the actual performance of parts with similar attributes. Finally, by determining the relationships of new parts to previously designed parts, it increases human productivity in the generation of new designs and the revision of old designs. The software necessary to implement design retrieval is the simplest of all GT software; it involves a simple query to a GT data base for specific feasible ranges of variables.

Computer-integrated manufacturing systems exemplify a more sophisticated and more profitable GT implementation. Such a computer-integrated manufacturing can be created by identifying a cluster of machines that can, or will be able to, service a particular family of parts. The computer-integrated manufacturing can then be streamlined to produce this part family optimally.

Purchasing support is a rather new GT application, yet almost every major manufacturing facility has a quasi-GT system, called a commodity code, already performing this task. A rigorous GT system may pay a tremendous financial reward by permitting all related parts, including those that do not obviously belong to the same families, to be identified throughout a factory or corporation. In one instance, a vendor of hoses offered a 50 percent reduction on hose prices if a corporation could identify all hoses and their attributes to be purchased over a given period. The corporation saved millions of dollars by rigorously following a GT system to identify all hoses to be ordered. A GT purchasing support system offers buyers a significant way to cut costs through knowledge and buyer leverage.

Service depot streamlining is a GT application which can help determine the most advantageous service parts stategy by identifying where alternate parts may be used. Standardization of parts in the service depot allows for substantial reduction in inventory and repair time, even if the standard is the most expensive item.

The premise of a GT system is that the similarity of parts and processes can be turned into substantial cost savings. The most far-reaching applications of GT will be made possible by the structuring of the parts data base itself. If the data base information is captured in attribute form and linked to applications by similarity, the ability of the data base to support manufacturing decision making will be greatly enhanced. This global application of GT to data base design is only now gaining popularity, and research on it is still in its infancy.

One of these far-reaching applications is in the area of process planning. By performing a rigorous analysis of manufacturing processes and parts to be made, a manufacturer will improve his ability to move from the present method of process planning into the more highly integrated future. For example, a manufacturer may evolve from his present variant process planner, which uses GT to match part families to process families, to a more sophisticated generative system in which more knowledge captured in the GT data base will be used to optimize the process plan. Eventually, a highly integrated system can be achieved that delays the final process planning step until the part is to be made, optimizing not only the process but also capacity utilization. Group technology, like network technology, will be a cornerstone of computer-integrated manufacturing systems.

Adaptive Closed-Loop Control

One difference between an automated system and an intelligent system is the amount and kind of feedback that is generated from an activity and passed up to a decision-making entity. This feedback allows a system to know its own state, to know when it is out of balance, and to respond to the imbalance until stability is achieved. This adaptive closed-loop control will be used at all levels of manufacturing systems from sensor-based feedback to robots or NC tools, to inspection station-based feedback to a cell controller, to a factory floor data collection system that feeds back to process planning and scheduling systems. This property of adaptive feedback is the key to improving product quality, with zero defects a realistic goal.

Adaptive feedback is also one key to better management, since only in this way can a manager know exactly the state of production, including exact costs. Systems of manufacturing have a feedforward property that will allow management to control the factory floor with an effectiveness and immediacy never before possible.

Feedback and feedforward properties can also provide machines and systems with self-diagnosis features. Thus, a manufacturing system can tell if something is wrong with it and what is wrong and can suggest the remedy to a higher entity. For example, researchers have demonstrated the ability to sense when a tool is about to break so that automated equipment can change the tool without the disruption caused by untimely failure. Next will be limited self-maintenance and repair capabilities. When a data-intensive system breaks down, the integrity of that data is threatened. Self-diagnosis will inform the data base system of the integrity of the data and, if it is threatened, the system will take either conservative or remedial action.

Factory Management and Control

The factory of the future will be managed and controlled through automated process planning, scheduling, modeling, and optimization systems. The successful implementation of large-scale factory-level systems depends upon structured analysis and design systems that depend heavily on GT. Limited structured analysis systems (see Figure 9–6), such as the Air Force-sponsored Integrated Computer-Aided Manufacturing Definition, have been in use for years in the analysis and design of large projects. Only through such systems can a manager know the exact state of his factory, and only through such exact knowledge of the present can a manager intelligently implement systems of manufacturing for the future. New systems development methodology packages, such as STRADIS, promise help in this area, but much work remains to be done before such systems are easily used by the actual decision makers. Similar work must be done on process planning and scheduling

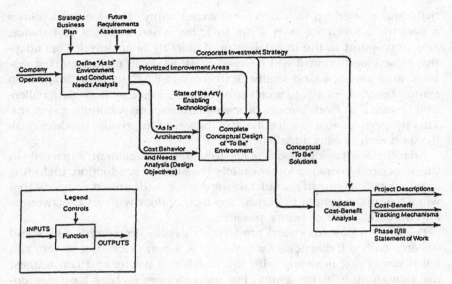

Figure 9-6. Example of U.S. Air Force Integrated Computer-Aided Manufacturing Definition of System Elements.

systems before they can use the feedback and feedforward properties of the hierarchical and adaptive closed-loop control systems to be found in future computer-integrated manufacturing systems.

Management functions will be hierarchically distributed so the "go" and "halt" decisions may be made effectively from many levels and by human or machine. Through the use of terminals on the factory floor and throughout the decision-making structure, the system can respond instantaneously to human command. At first, most of the decision making will rest in the hands of humans. Low level manufacturing systems now work in this way. As the systems become integrated at higher and higher levels, decision rules and methods will be built into them; systems will develop plans to carry out human-specified activities. On the authorized human's approval, the system will carry out the task, making low-level decisions on its own. If a low level decision-making entity does not have a certain level of confidence in its decision, it may pass the decision up to the next higher entity, be it human or computer. This new kind of man-machine interaction will allow humans to do what they do best: create, define, and communicate. The machine will do what it does best: work hard, steadily, and accurately.

Modeling and Optimization Systems

One tried and true method of representing an activity to a computer is through mathematical modeling. Computerized modeling tools, such as SLAM, have been used by simulation experts for years, but simulation is still more an art than a science. With experience and feedback, our ability to represent complex activities mathematically will be refined. A modeling and simulation package will be a necessary part of an intelligent structured analysis and design system. It is hard to overstate the importance of structured analysis and design systems; they will operate at high levels, with much built-in decision making.

System modeling will become a commonplace and necessary prerequisite to the successful design and implementation of large-scale manufacturing systems. This is because large projects contain too many facets to be managed effectively with only human memory and computation capability. Artificial intelligence techniques will be needed to reduce the tremendous amounts of data generated by such systems to humanly understandable terms. This intelligence must be of a higher order than the AI expert systems in existence today.

With the addition of AI, a modeling system can become an optimization system, guiding its human managers to the most productive, most cost effective, or highest quality utilization of resources. With such optimization capability, managers will be able to sit at their workstations and, in real time, analyze the various possibilities to determine optimal solutions and mixes. The availability of accurate information on cost, time, and quality will eliminate much of the guesswork in manufacturing decision making.

Computer-Integrated Manufacturing Systems

Computer-integrated manufacturing systems are expected to dominate the factory automation movement within ten years. Computer-integrated manufacturing will be tied into larger scale manufacturing systems, but it is valuable to consider the computer-integrated manufacturing as a critical unit or building block in total factory integration. A computer-integrated manufacturing may be described as an integrated system of machines, equipment, and work and tool transport apparatus, using adaptive closed-loop control and a common computer architecture to manufacture parts randomly from a select family. The hardware components of a computer-integrated manufacturing may include an NC tool, a robot, or an inspection station. The part family processed by the computer-integrated manufacturing is defined by GT classification. For greatest productivity, the computer integrated-manufacturing is optimized to produce only one family of parts, and conversely, the parts produced by the computer-integrated manufacturing are designed to facilitate processing by the computer-integrated manufacturing.

The concept of flexibility as used in a computer-integrated manufacturing includes:

• use of GT to achieve a part mix of related but different parts;
• batching, adding, and deleting of parts during operation;
• dynamic routing of parts to machines;
• rapid response to design changes;
• making production volume sensitive to immediate production of parts on demand; and
• dynamic reallocation of production resources in case of breakdown or bottleneck.

Computer-integrated manufacturing has been a reality in U.S. industry since its introduction in 1972, and the number of new FMS/computer-integrated manufacturing installations is doubling every two years. The number and flexibility of computer-integrated manufacturing is expected to increase, and the cost of computer-integrated manufacturing installations is expected to drop. Although the United States was largely responsible for the technological development of the computer-integrated manufacturing, Western Europe and Japan both have more computer-integrated manufacturing installations than this country. In fact, one of the most frequently cited computer-integrated manufacturing installations is located in the Messerschmidt-Boelkow-Blohm (MBB) plant in Augsburg, West Germany. The basic elements of this computer-integrated manufacturing are 25 NC machining centers and multispindle gantry and traveling-column machines; automated tool transport and tool changing systems; an AGV workpiece transfer system; and hierarchical computer control of all these elements. The computer-integrated manufacturing is used to build wing-carrythrough boxes for Tornado fighter-bombers. Comparisons by MBB of the performance of this computer-integrated manufacturing versus the projected performance of stand-alone NC machine tools doing the same work clearly show the advantages of the computer-integrated manufacturing approach:

• number of machine tools decreased 52.6 percent;
• workforce reduced 52.6 percent;
• tooling costs reduced 30 percent;

- throughput increased 25 percent;
- capital investment 10 percent less than for stand-alone equipment; and
- annual costs decreased 24 percent.

U.S. statistics for computer-integrated manufacturing installations are no less startling. A computer-integrated manufacturing at General Electric (GE) improves motor frame productivity 240 percent; an AVCO computer-integrated manufacturing enables 15 machines to do the work of 65; and at Mack Trucks, a computer-integrated manufacturing permits 5 people to do what 20 did before. In addition to productivity enhancements, the computer-integrated manufacturing offers increased floor space capacity. GE, for example, reported that floor space capacity was increased 50 percent, with a net floor space reduction of 30 percent. A computer-integrated manufacturing can make a factory more responsive to its market—GE reported a shortening of its manufacturing cycle from 16 days to 16 hours.

The technologies discussed above will be integrated to create computer-integrated manufacturing systems whose synergy will make the whole greater than the sum of its separate technologies. Computer-integrated manufacturing systems promise dramatic improvements in productivity, cost, quality, and cycle time. However, since full computer-integrated manufacturing has not yet been accomplished and depends on continued technological progress, the benefits are difficult to quantify accurately. Incremental gains from the implementation of individual technologies and subsystems will be substantial. These benefits are illustrated by the following data from five companies that have implemented advanced manufacturing technologies over the past 10 to 20 years:

Reduction in engineering design cost	15–30 percent
Reduction in overall lead-time	30–60 percent
Increase in product quality	2–5 times
Increase in capability of engineers	3–35 times
Increase in productivity of production operations	40–70 percent
Increase in productivity of capital equipment	2–3 times
Reduction in work in process	30–60 percent
Reduction in personnel costs	5–20 percent

The cumulative gains of total system integration can be expected to build on these results exponentially.

The long-range goal of computer-integrated manufacturing is the complete integration of all the elements of the manufacturing subsystems, starting with the conception and modeling of products and ending with shipment and servicing. It includes the tie-in with activities such as optimization, mathematical modeling, and scheduling.

A computer-integrated manufacturing system is created by the interconnection or integration of the processes of manufacturing with other processes or systems. The resultant aggregate system provides one or more of the following functions or characteristics:

- An information communication utility that accesses data from the constituent parts of the system and serves as an information communication and retrieval system.
- An information-sharing utility that integrates data across system elements into a unified data base.
- An analysis utility that provides a mathematical model of a real or hypothetical manufacturing system. Employing simulation and, when possible, optimization, this utility is used to characterize the behavior of the modeled system in various configurations.
- A resource-sharing utility that employs mathematical or heuristic algorithms to plan and control the allocation of a set of resources to meet a demand profile.
- A higher order entity that integrates information and processing functions into a more capable, effective processing system.

These functions and characteristics are not mutually exclusive in actual manufacturing systems; rather, they overlap significantly with all of the elements interconnected and integrated continually to form a single aggregated computer-integrated manufacturing. Perhaps the most important and least understood step in this process is the creation of an integrated system that is a higher order entity; this is the true system-building goal.

Both horizontal and vertical growth of computer-integrated manufacturing systems can be expected as the year 2000 approaches. State-of-the-art technology now includes small aggregates of computer-integrated tasks, often called islands of automation. Such islands of automation are found in design, where CAD workstations from different vendors share their data through a common data base and data conversion interfaces; in planning, with manufacturing resource planning (MRP) systems; and in production, where a work cell composed of a robot, machine tool, and inspection station may be coordinated by a cell controller.

In leading-edge plants, several of these islands of automation have been aggregated into larger manufacturing subsystems, termed continents of automation. At this level of integration, links exist between the design and engineering departments, with CAD terminals and data bases sharing data with CAE workstations and data bases. In planning, an MRP system can be linked to traditional data bases containing ordering and shipping information. On the factory floor, several work cells may be integrated with a material handling system to create a computer-integrated manufacturing.

In the factory of the future, these continents of automation will be integrated into worlds of automation that will eventually encompass not only entire factories, but also entire corporations. Because of the volume of data and complexity of decisions needed for full integration, a hierarchical structure is the only feasible way to achieve it.

A hierarchical structure has certain implications for the architecture of computer-integrated manufacturing systems. Data use and decision making must occur at the lowest levels possible. Only certain summary data will be passed upward in the hierarchy to be used in reporting the factory's state and in statistical trend analysis. Thus, an information and decision hierarchy is needed that practices management by exception. If additional information is required at upper levels, it will be requested. If local decision making cannot resolve a conflict, a decision will be requested from above. Conversely, management decisions may be communicated almost instantly throughout the system for rapid compliance.

The hierarchical structure further implies the use of distributed data bases and distributed processing. A mainframe computer may be the host computer to the factory of the future. Connected to it will be an array of minicomputers, one level down in the hierarchy, each acting as host controller to an intermediate level manufacturing system. A mix of local and centralized data storage will be appropriate for each computer. Below the minicomputers will be microcomputers acting as cell controllers, graphics workstations, or executive workstations.

The elements of this hierarchical structure can be thought of as subsystems, categorized by the role they play, although it must be remembered that categories may overlap considerably. Most subsystems of manufacturing fall into one of the following broad categories: (1) information and communication, (2) integration of processes, or (3) resource allocation. Note that each category cuts across traditional manufacturing boundaries. An information-communication subsystem, for example, could include a network that permits the geometric part data stored in the CAD data base to be transformed through GT techniques into an actual process plan and then into robot and NC programs communicated to the factory floor. The data would then be transformed and communicated for process scheduling and material handling, right up to the delivery of the finished product.

Information-oriented subsystems include traditional management information system and data processing roles. These subsystems will be able to expand to include geometrical data from CAD systems, material and process data from GT coding, parts-in-process data, and order and inventory data. Information subsystems will have analytic capabilities by which the data can be massaged for quality control and trend analysis. Data retrieval will be easier for operators and decision makers through the use of new query languages or programs that will allow nonexperts access to complex data. Most, if not all, manufacturing subsystems have strong information and communication functions, even if they are primarily process- or resource-oriented.

Manufacturing subsystems on the factory floor will integrate traditional manufacturing processes by coupling and controlling previously separated processes and by carrying out computer-generated process plans. At the lowest level, this will involve data communication from sensors to a computer-controlled machine or robot. This provides the real-time adaptive control necessary to improve the work quality and throughput of individual stations. At the next level, factory floor manufacturing subsystems can integrate several processes, such as an NC machining station, an automated inspection station, and the robot which services them. In this example, the coordination is supplied by a computer which controls the work cell. The process plan is downloaded from a computer, which may be in the CAE area, to the work cell controller, which coordinates the processing by the machines in its cell. With automated inspection and data collection, the process plan may be modified to eliminate defects by responding in real time to tolerance changes. At yet a higher level, work cells are integrated into a computer-integrated manufacturing so that an automated scheduling system can assign a part in process to the next available work cell that can perform the necessary operation. This allows a system with fewer parts in process, shorter lines, fewer holding areas, and much more efficient use of floor space.

Resource allocation subsystems span a broad scope from small-scale material handling systems serving individual work cells to broadly implemented systems that monitor and control inventory, schedule work, and allocate materials to the factory floor on tight schedules. Automated material handling systems can be integrated into work cells and families of work cells to produce a powerful computer-integrated manufacturing. In turn, the computer-integrated manufacturing can be linked to production planning and capacity planning systems to form full computer-integrated manufacturing systems.

One of the most important reasons for implementing small-scale subsystems of manufacturing now is that they can be successfully integrated into these higher order entities, computer-integrated manufacturing systems, that benefit from the synergy between operational programs, product data, and process data.

Summary

All of these advanced manufacturing technologies, from machine tools to the subsystems and computer-integrated manufacturing systems, provide the ability to perform traditional manufacturing tasks in a highly advantageous but nontraditional manner. Many of the individual technologies and subsystems of manufacturing can be implemented today and, in fact, must be implemented soon for a manufacturer to remain competitive. Real progress toward the factory of the future will take place through the higher level integration of these technologies. Although a handful of domestic manufacturers continue to make progress in implementing and integrating many of the technologies described in the Appendix, real barriers to full integration remain.

Specifically, standards are critically needed for the definition and communication of part data. At higher levels, the need is for proven systems of hierarchical control and feedback and usable methods of automated classification of parts and processes (GT). Required at the highest level are the evolution of structured analysis and design systems that include modeling and optimization packages, as well as intelligent user interfaces that can be used interactively by managers in real time. The technology is here now or just around the corner. U.S. manufacturing needs far-sighted management and trained manufacturing engineers to put the pieces together.

Chapter 10

Projected Trends
in CIM Technology

There are a number of directions in which CIM technology will proceed in the future, driven both by the need to overcome certain present obstacles and by the availability of concepts that come to practical fruition. The changes will affect the machines, the tools, the transport mechanisms, the control hardware and software, and the workpieces themselves.

Many of the trends in the following list are now in progress, but some will not be fully available for near-term application:

- Machine adjustable, tool changer compatible tools:
    cutting tools (boring bars, etc.);
    forming tools;
    gauges.
- Automatic tool loading/unloading into tool-changer magazines from tool crib.
- Automatic tool crib operation and connection to CIMs.
- Fully automatic tool exchanges between machines.
- Compact, large capacity, high-speed, random-access, tool changer magazines.
- Tool wear and breakage sensors and algorithms.
- On-line inspection.
- On-machine inspection:
    tool changer loading devices (e.g., robots);
    independent of tool storage, changer, and spindle.
- Adaptive error determination and compensation systems:
    drive axes, error profile compensation;
    alignment compensation;

    force and weight compensation;

    temperature compensation;

    wear compensation.

- Tighter coupling of CAD and CAM, e.g., CAE and CAD systems help designers create CIM-compatible designs.
- More comprehensive and data-accessible manufacturing monitoring systems.
- Comprehensive production planning and management systems to handle:

    inventory;

    ordering;

    batching definition;

    scheduling;

    routing;

    rescheduling/rerouting for contingencies.

- Design for automation approach stressing mechatronics.
- Fail-soft, graceful degradation of machines and systems.
- Improved reliability of CIM systems and sub-elements.
- Automatic generation of system initialization and start-up procedures, and a "checklist" for any stoppage or failure condition.
- Automatic part fixturing and defixturing, automatic part storage and retrieval.
- Automatic pallet and fixture storage and retrieval, with automatic insertion into and extraction from the system.
- Automatic identification and tracking of individual tools, pallets, fixtures, parts, carts, movable elements, etc., including automatic location during system start-up.
- Automated integration of more classes of manufacturing operations, including:

    milling;

    turning;

    forming;

    inspection;

    finishing;

    heat treating;

    shearing;

    assembly;

    testing.

- Improvements in automated chip flushing and clearing, part cleaning, chip collection and reclaiming, and coolant reconditioning.
- Improvements in automatic temperature control of parts, pallets, fixtures, machines, and coolants.
- Improvements in inherent machine tool and inspection machine ac-

curacy, as the increase in "near-net-shape" technology reduces the need for heavy machining and places the emphasis on high-precision machining.

- Better equipment maintainability through increased use of:
   automatic fault detection and isolation;
   built-in diagnostic and repair aids (test cooperative aids);
   modular construction;
   redundant equipment configurations (on a single machine tool allowing on-line repair).
- Increasingly larger CIMs (federations of CIMS) with better integration into the factory planning and operation systems (MRP, etc.).
- Move by world-class competitors toward unmanned factories.
- Concurrent design impact on manufacturing technology

Part IV

Competitiveness

Chapter 11

U.S. Industry Performance

The first generation of CIM/FMS in the United States goes back to the Sundstrand Machine Tool Company in 1965. However, it was not until 1970 that the Kearney & Trecker Company built the first elements of an FMS/CIM system. Despite the considerable interest CIM has generated, the application of CIM in the United States has spread slowly. We have identified nine U.S. companies that have sold systems. There are currently approximately 47 FMS/CIM systems in operation in the United States and several in planning or being installed.

Suppliers

CIM are not off-the-shelf turnkey operations. Each system is designed for a special purpose and the supplier must work very closely with the customer. Few companies have the ability or are willing to undertake building a CIM. However, there are many manufacturers that provide equipment that can be integrated as a component to the overall system. CIM is still a relatively new concept and buyers are concerned with future liability, failure to perform, etc. This requires the supplier to be well-capitalized to gain the potential buyer's confidence.

Suppliers and potential suppliers are machine tool builders, robot and material handling manufacturers, and control manufacturers. Currently, firms in the machine tool industry provide the overall responsibility for design, development, and installation of CIM. U.S. suppliers in each of the three industries will be briefly examined before describing current U.S. CIM suppliers.

Machine Tools

Machine tools are a central component of CIM, so machine tool builders are key suppliers. There are approximately 600 machine tool builders in the United States. A few are large, but most machine tool producers are relatively small. About two-thirds of the U.S. machine tool manufacturers employ fewer than 20 persons and less than 1 percent employ 1,000 persons or more. The 12 top producers made 85 percent of the machine tools produced in the United States in 1982. Table 11-1 lists the production of the top 12 machine tool builders for the years 1979 to 1983. Amca International and Oerlikon Buhrle are foreign firms which have acquired U.S. machine tool capacity (Giddings & Lewis and Motch & Merryweather, respectively).

The U.S. machine tool industry is extremely cyclical as is demonstrated in Table 11-2. U.S. domestic shipments fell 26.7 percent from 1981 to 1982, then dropped again, 50.0 percent, from 1982 to 1983 before beginning to recover in 1984. U.S. exports also fell, 8.9 percent from 1981 to 1982 and 42.3 percent from 1982 to 1983. Exports remained depressed through 1984. This partly reflects the U.S. economic recession, but also reflects decreasing worldwide competitiveness of U.S. machine tools. In 1977, the historic trend of machine tool trade surpluses was reversed. Since that time U.S. trade in machine tools has been in deficit. Imports took 37 percent of the market in 1983 and increased their share to nearly 40 percent in 1984. Import penetration has been most severe in technologies critical to CIM such as machining centers and NC lathes.

Table 11-1.   Top 12 U.S. Machine Tool Builders by Sales
(Millions of Dollars)

Builder	1979	1980	1981	1982	1983
Cincinnati Milacron	464	563	640	515	284
Litton	160	190	200	200	267
Textron	150	150	270	210	160
Cross & Trecker	267	320	368	343	150
Ex-Cell-O	230	290	280	235	145
Oerlikon Motch	–	–	–	217[2]	140
Giddings & Lewis	196	222	287	240[2]	135[2]
Bendix Corporation	285	475	400	300	130
Ingersoll Milling Machine	75	200	200	160	130
Lamb Technicon	175[1]	200[1]	275[1]	280	120
White Consolidated Industries	135	165	180	155	100
Esterline	–	–	–	–	98
Acme-Cleveland	200	257	252	210	–

[1]F. Joseph Lamb
[2]Amca International

Source: American Machinist, 1980, 81, 82, 83, 84.

Table 11-2. U.S. Machine Tool Shipments, Exports, Imports, Consumption, and Imports Share of Consumption

YEAR	SHIPMENTS	EXPORTS	IMPORTS	DOMESTIC CONSUMPTION	IMPORT SHARE
1967	$1,826	$225	$178	$1,780	10.0%
1968	1,723	217	164	1,670	9.8
1969	1,692	242	156	1,606	9.7
1970	1,552	292	132	1,392	9.5
1971	1,058	252	90	896	10.0
1972	1,269	238	114	1,145	10.0
1973	1,788	325	167	1,629	10.3
1974	2,166	411	271	2,026	13.4
1975	2,406	537	318	2,187	14.5
1976	2,178	515	318	1,982	16.0
1977	2,453	427	401	2,428	16.5
1978	3,143	533	715	3,325	21.5
1979	4,064	619	1,044	4,489	23.3
1980	4,812	734	1,260	5,338	23.6
1981	5,111	950	1,431	5,593	25.6
1982	3,749	575	1,218	4,392	27.7
1983	2,114	359	921	2,676	34.4
1984	2,428	383	1333	3,377	39.5

Source: U.S. Department of Commerce, Current Industrial Reports; Metalworking Machinery, MQ35W, IM-146, EM-522

Robotics

Over 50 firms produce robots in the United States. The International Trade Commission reports that the top six firms accounted for 80 percent of U.S. shipments in 1982. In 1983 and 1984, the top six firms accounted for only 65 percent of U.S. shipments.

Major producers in the United States are listed in Table 11-3. These robotics companies include major corporations with an existing high-technology emphasis such as General Electric, International Business Machines, Westinghouse; and venture-capital-funded firms spurred by innovation and the prospect of growth (for example, Automatix Inc., Advanced Robotics Corp., and Mobot); and established robot producers which either began in this field or entered robotics based on their machine tool/processing systems expertise (for example, Prab Robots and Cincinnati Milacron).

The largest user of industrial robots in the United States is the automotive industry. Other major users of robots are the aircraft, farm equipment, electrical equipment, and home appliance industries. It is expected that the use of robots will spread to other manufacturing industries as they become more cost-effective.

U.S. robot shipments, exports, imports, domestic consumption, and imports share of consumption from 1979 through 1983 are listed in Table 11-4.

Table 11-3. U.S.—Based Robot Vendors Sales (Millions of Dollars)

Company	1980	1981	1982	1983	1984	1985*	1986*
GMF	–	–	0.3	22.3	102.8	187.0	185.0
Cincinnati Milacron	29.0	50.0	32.0	51.0	52.5	61.0	65.0
Westinghouse/Unimation	40.0	68.0	63.0	36.0	44.5	45.0	40.0
Automatix	0.4	3.0	8.1	13.0	17.3	25.0	24.0
DeVilbiss	5.0	6.5	23.7	21.0	30.0	33.0	28.0
ASEA, Inc.	2.5	9.0	9.5	15.0	30.0	39.0	48.0
American Cimflex	–	–	–	–	–	18.0	32.0
IBM	–	–	4.5	9.0	12.0	16.0	NA
General Electric	–	–	1.8	11.0	10.0	13.0	NA
Prab Robots	5.5	8.2	12.5	13.5	11.0	12.0	NA
Intelledex	–	–	–	2.0	10.0	10.0	NA
Seiko	–	–	–	4.0	6.5	11.0	NA
GCA	–	–	1.5	6.0	13.5	34.0	NA
American Robot	–	–	–	2.8	8.0	20.0	NA
Cybotech	–	–	9.0	2.3	7.0	12.0	NA
Graco Robotics	–	–	–	5.0	9.3	20.0	NA
Advanced Robotics	1.7	0.8	6.6	3.2	6.0	NA	NA
ESAB	–	–	–	4.0	4.5	8.0	NA
Thermwood	–	1.0	2.5	2.0	4.5	NA	NA
Hobart	–	–	–	3.5	5.5	6.5	NA
Control Automation	–	–	–	0.5	3.0	NA	NA
U.S. Robots	–	–	1.0	1.7	2.5	NA	NA
Nordson	0.8	2.5	4.5	4.7	2.0	NA	NA
Mobot	0.8	0.6	1.3	1.6	1.6	NA	NA
Microbot	–	–	–	0.7	1.4	NA	NA
Precision Robots	–	–	–	0.3	1.0	NA	NA
Adept Technology	–	–	–	–	1.1	17.0	25.0
KUKA	–	–	–	–	–	30.0	25.0
Other	4.5	5.5	8.2	6.2	6.2	19.5	131.0
Total	$90.0	$155.0	$190.0	$240.0	$395.0	$595.0	$625.0

*Estimated.

Note: material designated NA is not available or is included in other category.

Source: CIM Newsletter, Prudential-Bache Securities, April 27, 1984; June 24, 1985; and May 23, 1986.

Table 11-4. U.S. Robot Shipments, Exports, Imports Consumption, and Imports Share of Consumption 1970 to 1983 (Millions of Dollars)

YEAR	SHIPMENTS	EXPORTS	IMPORTS	DOMESTIC CONSUMPTION	IMPORT SHARE
1979	$28.1	$ 8.9	$ 3.8	$ 22.9	16.4%
1980	64.1	20.8	4.3	47.5	8.9
1981	113.4	23.3	10.6	100.7	10.5
1982	142.8	20.3	15.1	137.6	11.0
1983*	168.7	33.7	28.9	163.9	17.7

*Estimated.

Source: U.S. International Trade Commission, *Competitive Position of U.S. Producers of Robotics in Domestic and World Markets,* December 1983.

Both the quantity and value of U.S. exports have increased each year, reaching an estimated 631 units valued at $33.7 million in 1983. For all years, U.S. exports of robots exceeded U.S. robot imports. The import penetration ratio has been rising since 1980 but was still below the 1979 ratio until 1983. Shipments, exports, imports, and apparent domestic consumption have all increased fairly rapidly over the five years, reflecting the general increase in the application of robots in industrial settings.

U.S. producers' domestic shipments of robots (as opposed to total shipments which are reported above) increased $19 million in 1979 to $122 million in 1982, and are estimated at $135 million in 1983. Shipments to the domestic market accounted for 85 percent of total U.S. producer shipments in 1982.

A large number of U.S. producers of robots, including most of the major vendors, are involved in agreements with major foreign robot firms. These agreements cover joint ventures, product or process licensing arrangements, marketing, distribution, and technology transfer. Because of these arrangements, dispersion of technology between countries has accelerated, and duplication of R&D efforts has undoubtedly decreased.

CAD/CAM

There were more than 70 CAD/CAM suppliers in the United States as of February 1984. In 1983, according to Merrill-Lynch, five vendors of CAD accounted for 72 percent of the U.S. market, as follows: IBM (23 percent), Computervision (22 percent), Integraph (11 percent), Calma (GE) (10 percent), and Applicon (6 percent). Other significant competitors include McDonnell-Douglas (McAuto), Auto-Trol Technology, and Gerber Scientific. The CAD/CAM market is growing and market shares have had a tendency to change fairly substantially from year to year.

The CAD installed base in 1981 and projected for 1985 and 1995 are given in the Table 11-5.

Table 11-5. U.S. CAD Installed Base (units)

Industry Group	1981	Proj. 1985	Proj. 1995
Mechanical Design	1,744	9,000	82,000
Electrical and Electronic	1,620	3,700	51,000
Civil Engineering/ Architecture	630	2,900	37,000
Mapping/Other	568	2,400	20,000
Total Installations	4,562	18,000	190,000

Source: Industry Week/Predicasts

U.S. shipments of CAD/CAM turnkey systems, including net exports, was $765 million in 1981. U.S. shipments of turnkey systems are predicted to increase to $2.75 billion by 1985 and $12.0 billion by 1995.

Projected growth rates are very high, due to expectations of declines in the price of CAD/CAM systems. In current dollars, the cost of an average turnkey CAD/CAM system is expected to fall from $400,000 in 1980 to $200,000 in 1990. These costs are for relatively large systems. Smaller stand-alone workstations based on microcomputers have been introduced in the $10,000 price range.

A 1987 study by Arthur D. Little projected a compound annual growth rate in CAD of about 30 percent from 1982 to 1992. The growth rate of CAM is expected to be lower, at 10 to 12 percent over this period.

The CAD/CAM market is currently dominated by U.S. turnkey suppliers with 80 percent of world sales. It is expected that there will be more fierce competition in the future from mini and mainframe computer manufacturers. In 1982, U.S. firms held nearly all of the domestic CAD/CAM market, 90 percent of the Western European market, and 70 percent of the Japanese market. Three European firms represented the most significant competition: Quest, Ferranti Cetec, and Racel-Redac. Prominent firms in Japan include Fujitsu, Hitachi, and Sharp.

FMS/CIM

Nine American companies have actually sold at least one FMS/CIM system. The nine companies with FMS/CIM sales and the companies to which the systems were sold are listed in Table 11–6. These nine are not responsible for all the systems in the United States. A few have been built by foreign companies and some have been developed by companies which have been using CNC machine tools for some time and have the in-house skills to interface their existing machines or new machines into a CIM. The companies that have done this are listed in Table 11–7.

The nine FMS/CIM suppliers are all large machine tool manufacturers. They are the companies one would expect to be in the forefront of the CIM market. They have the range of necessary skills and technology as well as the resources to develop CIM.

Some of the advantages reported by users integrating their own system are that the company can get the best machines and equipment to fit its needs from any source, and utilize its in-house expertise at minimal additional cost. On the other hand, a company putting its own system together could underestimate the complexity of the undertaking and risk spending considerable time and money getting the system running. We will continue to see companies building their own systems despite the

Table 11-6. U.S. CIM Suppliers with Sales or Self-Built Systems

SUPPLIER	USER
Cincinnati Milacron	FMC Corporation
	Vought Aero Products
	Cincinnati Milacron
	Plastics Machinery Group
	Caterpillar
	(Davenport)
	N.Y. Air Brake
	General Dynamics, Ft Worth
	General Electric, Evandale
Giddings & Lewis	General Electric
	Caterpillar (Aurora)
	Anderson Strathclyde
	McDonnell Douglas Astronautics
Ingersoll Milling Machine	J.I. Case
	Ingersoll Milling (3 systems)
Kearney & Trecker	Allis Chalmers
	Onan
	Avco-Lycoming, Williamsport
	Avco-Lycoming, Stratford
	Hughes Aircraft
	Rockwell International
	Mercury Marine
	Mack Truck
	Warner Ishi
	Cummins Engine
	Georgetown Manufacturing
	Deere & Company
	Sundstrand Aviation
White Sundstrand	Boeing Aerospace
	Buick, Detroit Diesel
	Watervliet Arsenal
	Caterpillar
Oerlikon-Motch	Rockwell Motch
Acme-Cleveland	Vickers
Mazak	Caterpillar, Peoria
	Mazak, Florence (2 systems)
Dearborn	Caterpillar, Decatur

Table 11-7. U.S. Manufacturers with
Self-Built Systems

Harris Press	1981
General Electric	1984
Pratt & Whitney	
General Dynamics Convair (3 systems)	1982
Westinghouse	1984

risks, particularly if, as some users perceive, U.S. CIM suppliers are not responsive to the needs of the customer.

Some of the larger manufacturers and multinationals have worked with machine tool builders to develop systems because they are planning to become CIM suppliers. General Electric has spent over $1 billion in the past five years putting together what it calls an "across the board factory automation capability." It has put together its process control, electric drive, and CNC products and acquired a CAD company, a microchip maker, and several robot manufacturing licenses. Westinghouse is concentrating on developing CIM to process sheet metal. McDonnell Automation (McAuto) had a major exhibition at the International Machine Tool Show in Chicago, offering its services as a factory automation supplier. Other leading electronics companies such as Hewlett-Packard, Digital Equipment, and IBM may also look into developing marketable integrated systems. With the substantial resources these large multinationals have, they could present a real challenge to the traditional machine tool builders in the CIM market.

So far, the largest multinationals have shown no interest in developing their own machine tools or in acquiring any machine tool companies. Instead, they have been concentrating on the ancillary products that go in CIM, particularly software. The market for ancillary products may become more lucrative than the market for machine tools. If these multinationals do become strong participants in the CIM market, from where they source the machine tools will have a direct impact on the U.S. machine tool industry.

Smaller companies with innovative computer capabilities or expertise in a narrow high technology product line related to CIM are beginning to have an effect on the CIM market. One of the main drawbacks of CIM today is software. Companies that come up with innovative software packages and machine controls should do well in the market.

Allen-Bradley and General Electric, for example, believe that machine tool builders which now make their own CNCs will not be able to keep up with the specialist manufacturers in terms of product improvement and price reduction, and so they will start buying from specialists.

Products such as systems monitors, terminals, and communication devices are prime targets for specialists. Whether those companies specializing in controls or software actually get in the business of putting systems together for a customer remains to be seen. Since this would require resources beyond the means of most of these companies, the more likely outcome will be that these smaller, specialized companies, will push for standards and design their equipment to be compatible with systems, or sign agreements with systems builders.

Some firms are finding that the most efficient route to gaining access to additional skills and product lines is to pursue joint ventures. Joint ventures are common among companies trying to reposition themselves strategically. Examples include Acme-Cleveland-Mitsubishi, Westinghouse-Mitsutoki, and GM-Fanuc.

Most of these joint ventures have offered the potential for low cost, reliable overseas manufacturing and marketing for the U.S. partner, and an enhanced marketing network in this country for the foreign partner. They represent the trend toward centralization taking place in the industry. These joint ventures raise some concerns as to the effect they will have on the long-run competitiveness of machine tool manufacturing facilities located in the United States. If the practice of overseas procurement or production by U.S. companies of machine tools for sale in the United States were to become widespread, there would be the long-term danger that U.S. companies would end up more as distributor channels for foreign-built machine tools than as manufacturers in this country.

Users

In the United States 47 FMS/CIM systems have been identified to be in operation. This may seem a low number, especially since numerous articles about CIM and its application give the impression that there is an explosion of CIM.

Table 11–8 is a list of FMS/CIM users and the suppliers of the systems. Figure 11–1 shows the number of systems in operation in the United States by year.

From the list of users it is apparent that the farm machinery, automobile, aircraft, and locomotive industries are the pioneers of CIM in the United States. These are generally large companies with the necessary capital and are under heavy pressure from foreign competition. These are the prime candidates for CIM based on their product mix, size of part batches, families of parts, etc.

Foreign Penetration of the U.S. Market

Of the 47 CIM in operation in the United States, five were built by foreign companies, an import penetration rate of slightly less than 11 percent. Comau of Italy, Mazak of Japan, and Shin Nippon Koki of Japan are the major foreign competitors in the U.S. market at this time. That few foreign built systems have been sold in the United States may seem to

Table 11-8.   FMS/CIM in the United States

USER	SUPPLIER
Vought Aero Products	Cincinnati Milacron
FMC Corporation	
Cincinnati Milacron, Plastics	
General Dynamics Convair (3 systems)	
General Dynamics, Fort Worth	
General Electric, Evandale	
Caterpillar, Davenport	
New York Brake	
Caterpillar Aurora (2 systems)	Giddings & Lewis
General Electric, Erie	
McDonnell Douglas	
J.I. Case, Racine	Ingersoll Milling
Ingersoll Milling, Rockford (3 systems)	
Allis-Chalmers	Kearney Trecker
Deere & Company	
Avco-Lycoming, Williamsport	
Avco-Lycoming, Stratford	
Rockwell-International, Newark	
Hughes Aircraft, El Segundo	
Mack Truck, Hagerstown	
Onan, Minneapolis	
Mercury Marine, Fond du Lac	
Georgetown Manufacturing	
Cummins Engine	
Sundstrand Aviation	
Warner Ishi, Shelbyville	
Boeing Aerospace	White Sundstrand
Boeing Aerospace	Shin Nippon Koki
Watervliet Arsenal	
Buick, Detroit Allison	
Caterpillar	
Vickers	Acme-Cleveland
Harris Press, Ft. Worth	Harris Press
Borg Warner	Comau
Buick Gear & Axle	
Caterpillar, Peoria	Mazak
Mazak, Florence (2 systems)	
Rockwell Motch	Oerlikon-Motch
Caterpillar, Decatur	Dearborn
Pratt & Whitney	Pratt & Whitney
Westinghouse	Westinghouse

indicate that U.S. suppliers have a firm hold on the U.S. CIM market. This could also be rather deceptive. Some foreign companies are developing considerable CIM experience in their own countries and then moving to the United States to get into the market. At the International Machine Tool Show (IMTS) in Chicago, the major emphasis was on the modular approach to CIM development and technologies. Many foreign companies offered CIM development including: Hitachi-Seiki, LeBlond-Makino (which had a system operating at IMTS), Toshiba, OKK, Toyoda, and SNK from Japan; and Scharmann and EMAG/UMA Corporation of West Germany.

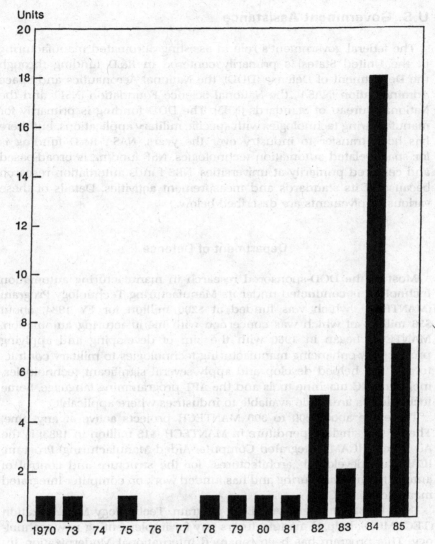

Figure 11-1. FMS/CIM Demand in the United States by Year

Some of the bigger foreign companies are making direct investments in the United States to establish a presence while others are just setting up service centers. Since the supplier and the user must work very closely together for the planning, installation, and debugging stages of the CIM project, foreign companies that do not have investments in the United States will be at a disadvantage.

U.S. Government Assistance

The federal government's role in assisting automated manufacturing in the United States is primarily centered in R&D funding through the Department of Defense (DOD), the National Aeronautics and Space Administration (NASA), the National Science Foundation (NSF), and the National Bureau of Standards (NBS). The DOD funding is primarily for manufacturing technologies with specific military applications, but there has been transfer to industry over the years. NASA R&D funding is for space-related automation technologies. NSF funding is broad-based and centered primarily at universities. NBS funds automation research because of its standards and measurement activities. Details of these various involvements are described below.

Department of Defense

Most of the DOD-sponsored research in manufacturing automation technology is conducted under its Manufacturing Technology Program (MANTECH), which was funded at $200 million for FY 1984, about $56 million of which was concerned with manufacturing automation. MANTECH began in 1960 with the aim of developing and applying productivity-enhancing manufacturing technologies to military contractors. It has helped develop and apply several significant technologies, including NC machine tools and the APT programming language. Some technologies are made available to industries where applicable.

There are about 200 to 300 MANTECH projects active at any time. The largest single expenditure in MANTECH ($18 million in 1983) is the Air Forces' ICAM (Integrated Computer-Aided Manufacturing) Program. ICAM has developed "architectures" for the structure and control of automated manufacturing and has funded work on computer-integrated manufacturing (CIM).

Until FY 82, another MANTECH program, Technology Modernization (TECHMOD), helped manufacturers pay for implementing new technology. This program has been renamed International Modernization Incentives Program (IMIP) and separated from MANTECH funding. IMIP is used to supplement cost-reimbursable contracts with DOD.

In addition to the MANTECH programs, two other DOD agencies fund longer term, more basic research in manufacturing automation technologies. The Defense Advanced Research Projects Agency (DARPA) funds research in robotics, sensory control, and artificial intelligence. The Office of Naval Research (ONR) mainly supports university-based research in precision engineering as well as other topics.

National Aeronautics and Space Administration

NASA supports research in manufacturing automation, with funding of $5.9 million in FY 1984. One area of research is robotics and manipulators for applications on space missions. Another area is the development of advanced computer-aided engineering systems and linkages with data management software packages.

National Science Foundation

NSF funds several programs with application to manufacturing automation, with a budget of about $7 to $9 million in FY 1984. The various areas of attention include discrete manufacturing, CAD, CAM, computer-aided testing, university research centers in robotics and in materials handling, touch and vision sensors, robot programming languages, computer architectures, and control systems.

National Bureau of Standards

The NBS Center for Manufacturing Engineering conducts research related to manufacturing automation with a budget of about $7.5 million in FY 1984. Among NBS's successes has been the development of a set of standards called IGES (Initial Graphic Exchange Specification) which enables different graphics systems to communicate with one another. The Automated Manufacturing Research Facility (AMRF) at NBS has been developed to serve as a laboratory for various kinds of automation research including CIM. Research will be primarily in the areas of measurement technology and interface standards.

Outlook for U.S. CIM Market

There are basically three ways in which a company can get a CIM. A company can buy the entire system from a domestic or foreign supplier who makes virtually all the machines and equipment, a company can use its existing NC machine tools or buy new machines and related equipment and develop the system itself, or a company can contract out to a company that buys the machine tools and related equipment and does the integrating and installation.

The first method is the most common and will continue to be over the next few years. The six CIM suppliers that are currently leaders in the U.S. market will continue to lead although they will be challenged by foreign CIM suppliers and possibly other U.S. machine tool builders that decide to get into the market. However, it will be difficult for other U.S. machine tool builders to get into the CIM market unless it begins to expand more rapidly. A potential user is more apt to buy a system from an experienced supplier and the suppliers with experience should be able to underprice new machine tool builders coming into the market since they are much further along the learning curve and may be able to install the system more efficiently. The major competition for current U.S. CIM suppliers for the next three to five years will come from foreign suppliers. Foreign suppliers are not yet firmly established in the U.S. CIM market; however, there are strong indications of their intent to do so.

Certain large multinationals initially envisioned becoming suppliers of the complete factory of the future. Demand for this technology has failed to materialize, forcing these multinationals to revise their strategies to supply smaller systems. This has brought them into direct competition with established machine tool suppliers. Indications are that the machine tool suppliers will continue to dominate the CIM market for at least the next three to five years.

The second market, building your own system, is being used by some companies and will continue to be used. For companies that want to retrofit existing machines or buy new machines of their choice, building their own system, provided they have the software skills, is a viable alternative. Standardization will have offsetting effects on use of this method. More standardization, and the willingness of suppliers to use any machine will reduce use of this approach. However, increased standardization will reduce the level of skills required for in-house user development and installation.

The third method, contracting out to a company that designs the system, buys the machines, and provides software and installation, may have the greatest potential. Increased market share will go to companies that can write the software and build the controls necessary to link the machines and equipment together. Standardization in communication between system components will promote this method as well as those discussed above.

Chapter 12

Assessment of World-Class Competition

It is difficult to make a factual determination on the competitiveness of CIM producers in one country versus those in another because producers within a country can vary considerably in ability; quantifiable data that forms a basis for comparison is generally proprietary, and qualitative measures can be subjective. Competitiveness depends on a number of factors. Each of these factors will be considered separately followed by an overall assessment.

Trade Trends

As has been pointed out, there are no trade data for CIM since they do not neatly fit into any Standard Industrial Classification (SIC) or other classification system. In the absence of trade data, comparisons and trends must be based on systems identified around the world. In this study, more than 200 CIM systems have been identified. Table 13–1 shows how many CIM systems there are in each country and how many were imported into each country.

From Table 13–1, it is apparent that trade in CIM at this time is not significant. Since this is a new market, however, the trends are important. The biggest CIM markets in non-Communist countries today are in Japan, the United States, Italy, West Germany, the United Kingdom, and Sweden. The United States has few foreign systems and should be expected to represent a large potential future market. It will be the target of foreign firms, particularly Japanese, Italian, and West German.

There are several possible explanations for the current low volume of trade in CIM, some of which are examined in the discussion of competitive factors below.

Table 12-1   Number of FMS/CIM Systems by Country and FMS/CIM Imports/Exports

Country	# of Systems	# of Imports	# of Exports
United States	47	5	2
Japan	50	0	6
Italy	37	0	14
West Germany	25	0	2
United Kingdom	15	4	0
Sweden	14	N/A	N/A
The Netherlands	0	N/A	N/A
Switzerland	2	0	0
France	7	5	1
USSR	2	2	0
East Germany	5	1	0
Korea	1	1	N/A
Taiwan	1	0	N/A
China	0	0	N/A
TOTAL	206	18	25

Prices

An important consideration, and one of the leading barriers to CIM, is the price of CIM. Each CIM system is unique so prices vary tremendously from system to system. In addition, for a true price comparison, only considering the up-front expense is not enough. The full installation costs, maintenance, durability, and reliability must also figure into costs.

Japanese machine-tool builders have been able to sell machine tools in the United States for 10 to 40 percent below U.S. producers' prices. Part of this difference is due to the high dollar/yen exchange rate but the Japanese have lower material costs, and reportedly superior machine tool manufacturing facilities. Since machine tools are generally the big ticket hardware items in CIM, this cost advantage helps make the foreign competitors' CIM system the least expensive in terms of up-front costs.

Performance Features and Quality

It is difficult to determine which systems have the best performance and are the most reliable because there are so few foreign systems in the United States for comparison, and even if there were, this information is not generally disclosed candidly. Nevertheless, based on studies of the machine tool industry, the robotics industry, the CAD/CAM industry, and interviews with CIM users and suppliers, a general picture can be drawn. U.S. competitiveness in each industry will be briefly examined before moving on to CIM.

Machine Tools

In a 1982 survey, U.S. purchasers of both U.S. made and foreign made machine tools were asked to rate producers regarding the engineering of their products. Users rated U.S. producers only slightly higher than Japanese producers. When the machine tool categories were broken down into types of machine tools used, U.S. products were rated first and Japanese products second in the metal-cutting categories, and vice versa in the metal-forming category. According to this survey, U.S. producers have a lead in large, sophisticated numerically controlled machine tools for use in production of aircraft, military equipment, and other specialized products. This is in line with a view that U.S. machine tools and CIM are most rigid and therefore more capable of heavy cuts. Of the European suppliers, the West Germans most nearly parallel the U.S. machine tools in construction.

The Yano report says that for such items as spindle speed, maximum allowable torque, spindle motor power, and cutting efficiency (quantities of cutting chips), the United States is ahead of Europe and particularly Japan.

This may be more a function of Japanese targeting rather than differences in technology. Japanese companies mainly manufacture small and medium-sized machining centers for general machining, while U.S. and West German companies mainly manufacture large and more powerful machine tools. The Japanese appear to hold a leadership position in designing and building CIM for small to medium prismatic parts and for rotational parts.

Computer Hardware/Software

The electronic content in CIM, both in hardware (sensors, process controllers, CAD systems, and inspection devices) and in software, is becoming larger and more sophisticated, making it difficult for traditional machine tool builders to move into the CIM business. It is one area, however, where the United States is widely recognized as the world leader. Japan lags behind the United States in this area.

Despite developments, software is considered the largest single problem area in the application of CIM. The unique aspects of each CIM make it difficult for CIM suppliers to use software packages from previous CIM systems as a guide for a newly designed system.

As the scope of CIM expands from machining to assembly, the importance and complexity of software will increase. This bodes well for U.S. suppliers.

Industrial Robots

Japan is farther ahead in the application of industrial robots in CIM. In terms of technology, Japan, the United States, and West Germany are relatively equal.

Availability

A CIM system is not bought off a showroom floor so immediate availability and delivery is not as important a factor as it is for stand-alone machine tools. The process of designing and installing a CIM system can range from six months to several years. The CIM supplier that has the experience and can design and install CIM has significant advantage over his competitor.

If CIM turns into standard, turnkey operations, availability will become more important. Some companies in Japan claim to have modular "CIM" or "CIM kits." These are probably small FMC systems. At this point, availability is a neutral issue and does not favor suppliers from any country. Japan has been able to provide some system elements faster than U.S. organizations.

Supplier/User Relationships

Suppliers and users must work closely together over an extended period covering the designing, installing, and training phases of the project. Clearly language barriers (Japan–U.S.), and lack of accessibility to each other could be problems. So far, most of the CIM systems in operation today have been built by companies in the same country. This trend is likely to change over the next five years. In order for a company to build a system in another country, it is desirable to have some form of permanent representation in that country. This is already taking place with U.S., Japanese, West German, and Italian machine tool builders and CIM suppliers locating all over the world.

Foreign companies with a subsidiary in the United States are beginning to make inroads in terms of service and responsiveness to customers. Based on discussions with U.S. CIM users, some feel that U.S. suppliers are not as responsive to the needs of customers, or as committed to making the system work as foreign suppliers. These same people expressed frustration because they feel that U.S. equipment is as good or better than foreign equipment.

During interviews, U.S. CIM users were asked about the purchase of a foreign system. Most users did not consider foreign suppliers. The off-road and agricultural heavy machinery builders are two exceptions. Overall, the decision process in the purchase of a CIM system appears to favor the domestic builder. The complexities associated with development and installation of a CIM system lead to the requirement of a close working relationship where the builder and user can react quickly to problems during planning, installation, and operation of the system. U.S. CIM users did not envision this type of relationship with a foreign CIM supplier. However, international competitors have quickly attacked new opportunities for business by responding to all of the user's needs.

Marketing and Distribution

In the 1970s U.S. CIM suppliers had a corner on the CIM market. During this period, marketing was primarily aimed at educating potential U.S. CIM users about the benefits of systems. While U.S. CIM suppliers were selling the concept to potential customers, companies in other countries, many with significant government assistance, were learning how to build CIM systems. Today, there are experienced CIM suppliers in at least seven countries.

These suppliers have become established in their own market and are now entering foreign markets, the United States being a prime target. This competition highlights the importance to U.S. firms of aggressive marketing and establishing good distribution channels.

Effective marketing influences the perceptions users have about a product. The Japanese have been very successful in creating the perception that they are in the forefront of new CIM technology and in the application of CIM. In fact, several representatives from different U.S. CIM suppliers expressed dismay that the Japanese were getting so much attention for their advances in technology when the technology had been developed by U.S. companies several years earlier. It does appear, however, that the systems at Mori Seiki, Iga, Yamazaki, Mino-Kamo, and Fanuc, Fugi are complex, state-of-the-art facilities.

Capital Availability

The cost of capital affects competitiveness in two ways. It can affect demand by raising or lowering the cost of borrowing to purchase. A customer is unlikely to use retained earnings to purchase high priced CIM system. The cost of capital also can raise the price the supplier must charge in order to be profitable if the producer must borrow money to deliver the product. The domestic market is affected in both ways. Foreign markets are affected by the product price effect, but not usually by the cost of borrowing to purchase.

Japanese CIM suppliers, while not yet very successful in penetrating the U.S. CIM market, have a great price advantage due to the strength of the U.S. dollar in relation to the Japanese yen and lower real interest rates. Japanese machine tools suppliers have certainly capitalized on the high value of the U.S. dollar and can be expected to press their advantage in the U.S. CIM market.

Overall Competitiveness

CIM is a precursor to the mechatronic factory of the future. It draws on technology from three important industries: machine tools, robotics and material handling, and electronics. The United States is not behind, technologically, in any of these three fields, yet it is experiencing a decline in the competitiveness of its machine tool industry and is facing serious competition in the robotics industry. The computer software industry is one area in which the United States has the lead. As it turns out, computer software is a key ingredient in CIM. This gives U.S. CIM suppliers one big advantage. However, after examining the various factors that affect competitiveness it is clear that no one country's industry has a clear advantage. The United States, Japan, Italy, Sweden, and West Germany are all developing a commercial capability. The United States' industry has been responsible for virtually all the technological innovations incorporated in CIM but has been unable to translate that technology into a commercial advantage.

U.S. CIM producers have been able to maintain their domination of the domestic CIM market, but they have not been particularly successful at exporting CIM. On the contrary, there are signs that U.S. CIM producers will have their hands full protecting the U.S. market from foreign penetration. It is too early to predict the fate of U.S. CIM suppliers. They have the ability to stay competitive, and even move ahead of the competition, if they can capitalize on the U.S. edge in computer software. CIM users generally agree that U.S. equipment is somewhat superior, but add that U.S. CIM suppliers are becoming less competitive. Why?

- Service and responsiveness of suppliers is most important. There is overwhelming agreement that U.S. suppliers are not as responsive to the needs of the customers, not as committed to making the system work, and are not able to deliver orders as scheduled.
- There is a general feeling that U.S. CIM suppliers lack the hands-on experience of running a CIM system. Most of the Japanese CIM suppliers developed a CIM system for their internal use before marketing the product. Some U.S. CIM suppliers do have CIM systems in their own factories but this is after much experimenting on the floors of customers.
- There is a general feeling that foreign companies are putting more money into research and development, thereby advancing CIM technology rapidly, although not yet surpassing the United States.

U.S. CIM suppliers have benefitted from the tendency of U.S. CIM users to give them preference. As other countries develop experience in CIM, and can offer comparable equipment at equal or lower prices along with equal or superior service, potential CIM customers may turn to foreign suppliers, especially if U.S. firms do not begin to develop better relations with users.

The competitiveness of the U.S. industry also depends on the rapidity with which demand for CIM increases in the United States. Currently, in the U.S. market; U.S. suppliers have an advantage. If demand grows fast enough to allow them to move down the learning curve more quickly than competitors (some of which have smaller domestic market bases) the U.S. industry will be in a good position. Although the factors affecting demand are very complicated, one of them is confidence in the technology and this factor is closely tied to confidence in U.S. suppliers.

Scenarios for Market Development

The key question for future competitiveness in CIM is the speed and extent to which U.S. manufacturers move to automate their factories. There are three scenarios that could develop:

1. Demand for CIM in the United States grows moderately or rapidly over the next five to seven years. U.S. CIM suppliers become stronger with increased demand. U.S. manufacturers (users) become more

competitive and reverse the trend of increased movement offshore of capital goods industries.

2. Demand for CIM in the United States grows modestly over the next five to seven years. Foreign CIM suppliers, using the experience gained in their own markets and their price advantage, undercut U.S. CIM suppliers and take a big share of the market. U.S. manufacturers (users) become more competitive using foreign-supplied systems.

3. Demand for CIM in the United States stagnates. U.S. CIM suppliers have no market. U.S. CIM and other manufacturers become less competitive with foreign competition.

The first scenario is the best for the U.S. and could be within reach, if our approach to CIM changes significantly. The second scenario, while good for U.S. manufacturing industries and the economy as a whole, is not good for U.S. CIM suppliers and weakens U.S. national security. The third scenario creates domestic economic problems.

The critical issue in all three scenarios is the speed and extent to which U.S. manufacturers automate their factories. To some extent, the U.S. government can influence the eventual outcome, with a change in policy that protects the U.S. industrial production base.

Points of Contact

CIM/CAD/CAM Organizations

Manufacturers

Adage, Inc.
1 Fortune Dr.
Billeria, MA 01821
(617) 667-7070

Applicon, Inc.
32 Second Ave.
Burlington, MA 01803
(617) 272-7070

Auto-trol Technology Corp.
12500 N. Washington St.
Denver, CO 80233

Avera Corp
200 Technology Cir.
Scotts Valley, CA 95066
(408) 438-1401

CADAM, Inc.
1935 N. Buena Vista
Burbank, CA 91504
(213) 841-9470

CAD/CAM, Inc.
2844 East River Rd.
Dayton, OH 45439
(513) 293-3381

Cadlinc, Inc.
700 Nicholas Blvd.
Elk Grove Village, IL 60007
(312) 228-7300

CGX Corp.
42 Nagog Park
Alton, MA 01720

Cincinnati Milacron
Machine Tool Div.
4701 Marburg Ave.
Cincinnati, OH 45209
(513) 841-8100

Computervision Corp.
201 Burlington Rd.
Bedford, MA 01730
(617) 275-1800

Control Data Corp.
Manufacturing Industry Marketing
P.O. Box 0
Minneapolis, MN 55440
(612) 835-8100

Design Aids, Inc.
27822 El Lazo Blvd.
Laguna Niguel, CA 92677
(714) 831-5611

GE Calma Co.
5155 Old Ironsides Dr.
Santa Clara, CA 95050
(408) 727-0121

Gerber Scientific Instrument (GSI)
83 Gerber Road W.
South Windsor, CT 06074
(203) 644-1551

Gerber Systems Technology, Inc.
40 Gerber Rd. E.
South Windsor, CT 06074
(203) 644-2581

Giddings and Lewis
305 W. Delavan Dr.
Jamesville, WI 53547
(608) 756-2363

Graphics Technology Corp.
1777 Conestoga St.
Boulder, CO 80301
(303) 449-1138

IBM Corp.
1133 Westchester Ave.
White Plains, NY 10604
(914) 696-1960

Intergraph Corp.
One Madison Industrial Park
Huntsville, AL 35807
(205) 772-3411

Kearney and Trecker Corp.
West Allis, WI 53214

Manufacturing and Consulting
    Services, Inc.
2960 South Daimler Ave.
Santa Ana, CA 92705
(714) 540-3921

The Manufacturing Productivity
    Center
IIT Research Institute
10 W. 35th St.
Chicago, IL 60616
(312) 567-4800

Manufacturing Software and Services
Div. of LeBlond Makino Machine
    Tool
7667 Wooster Pike
Cincinnati, OH 45227

Matra Datavision, Inc.
Corporate Place I
99 South Bedford St.
Burlington, MA 01803

McDonnell-Douglas Automation Co.
P.O. Box 516
St. Louis, MO 63166
(314) 233-2299

Mentor Graphics Corp.
10200 S.W. Nimbus Ave., G7
Portland, OR 97223
(503) 620-9817

Prime Computer, Inc.
1 Speen St.
Framingham, MA 01701
(617) 872-4770

Spectragraphics Corp.
3333 Camino Del Rio S.
San Diego, CA 92108
(714) 584-1822

St. Onge, Ruff & Associates, Inc.
617 W. Market St.
P.O. Box 309M
York, PA 17405
(717) 854-3061

Summagraphics Corp.
35 Brentwood Ave.
P.O. Box 781
Fairfield, CT 06430
(203) 384-1344

Summit CAD Corp.
5222 FM 1960 W. 102
Houston, TX 77069
(713) 440-1468

Synercom Technology, Inc.
500 Corporate Dr.
Sugar Land, TX 77478
(713) 491-5000

Technology Research Corp.
8328-A Traford Ln.
Springfield, VA 22152
(703) 451-8830

Tektronix, Inc.
P.O. Box 500
Beverton, OR 97077
(503) 682-3411

Telesis Corp.
21 Alpha Rd.
Chelmsford, MA 01824
(617) 256-2300

T&W Systems, Inc.
7372 Prince Dr., No. 106
Huntington Beach, CA 92647
(714) 847-9960

Valid Logic Systems, Inc.
650 North Mary Ave.
Sunnyvale, CA 94086
(408) 773-1300

VG Systems, Inc.
21300 Oxnard St.
Woodland Hills, CA 91367
(213) 346-3410

Robot Manufacturers

Accumatic Machinery Corporation
3537 Hill Ave.
Toledo, OH 43607
(419) 535-7997

Acrobe Positioning Systems Inc.
3219 Doolittle Dr.
Northlake, IL 60062
(312) 273-4302

Anorad Corporation
110 Oser Ave.
Hauppauge, NY 11788
(516) 231-1990

Armax Robotics, Inc.
38700 Grand River Ave.
Farmington Hills, MI 48018
(313) 478-9330

Automatix Incorporated
217 Middlesex Turnpike
Burlington, MA 01803
(617) 273-4340

Automaton Corporation
23996 Freeway Park Dr.
Farmington Hills, MI 48024
(313) 471-0554

Binks Manufacturing Company
9201 W. Belmont Ave.
Franklin Park, IL 60131
(312) 671-3000

Cincinnati Milacron
215 S. West St.
South Lebanon, OH 45036
(513) 932-4400

Comet Welding Systems
900 Nicholas Blvd.
Elk Grove Village, IL 60007
(312) 956-0126

Control Automation, Inc.
P.O. Box 2304
Princeton, NJ 08540
(609) 799-6026

Cybotech Corporation
P.O. Box 88514
Indianapolis, IN 46208
(317) 292-7440

The DeVilbiss Company
300 Phillips Ave.
P.O. Box 913
Toledo, OH 43692
(419) 470-2169

General Electric Co.
Automatic Systems
1285 Boston Ave.
Bridgeport, CT 06602
(203) 382-2876

General Numeric Corporation
390 Kent Ave.
Elk Grove Village, IL 60007
(312) 640-1595

GMFanuc Robotics Corp.
5600 New King St.
Troy, MI 48098
(313) 641-4100

Graco Robotics Inc.
12898 Westmore Ave.
Livonia, MI 48150
(313) 261-3270

Hitachi America Ltd.
59 Route 17-S
Allendale, NJ 07401
(201) 825-8000

Hobart Brothers Company
600 W. Main St.
Troy, OH 45473
(513) 339-6011

IBM
P.O. Box 1328
Boca Raton, FL 33432
(305) 998-2000

Industrial Automates Inc.
6123 W. Mitchell St.
Milwaukee, WI 53214
(414) 327-5656

Intelledex
33840 Eastgate Circle
Corvallis, OR 97333
(503) 758-4700

International Intelligence/Robomation
6353 El Camino Real
Carlsbad, CA 92008
(714) 438-4424

I.S.I. Manufacturing, Inc.
31915 Groesbeck Highway
Fraser, MI 48026
(313) 294-9500

Lamson Corporation
P.O. Box 4857
Syracuse, NY 13221
(315) 432-5500

Mack Corporation
3695 East Industrial Dr.
Glagstaff, AZ 86001
(602) 526-1120

Nordson Corporation
555 Jackson St.
Amherst, OH
(216) 988-9411

Nova Robotics
262 Prestige Park Rd.
East Hartford, CT 06108
(203) 528-9861

Pickomatic Systems
37950 Commerce
Sterling Heights, MI 48077
(313) 939-9320

Positech Corporation
Rush Lake Road
Laurens, IA 50554
(712) 845-4548

Prab Robots, Inc.
5944 E. Kilgore Road
Kalamazoo, MI 49003
(616) 349-8761

Reis Machines
1426 Davis Road
Elgin, IL 60120
(312) 741-9500

Rob-Con Ltd.
12001 Globe
Livonia, MI 48150
(313) 591-0300

Sandhu Machine Design Inc.
308 S. State St.
Champaign, IL 61820
(217) 352-8485

Schrader Bellows/Scovill Inc.
200 W. Exchange St.
Akron, OH 44309
(216) 375-5202

Seiko Instruments, Inc.
2990 W. Lomita Blvd.
Torrance, CA 90505
(213) 330-8777

Sigma
6505C Serrano Ave.
Annaheim, CA 92807
(714) 974-0166

Sormel/Black & Webster
281 Winter St.
Waltham, MA 02254
(617) 890-9100

Sterling Detroit Company
261 E. Goldengate Avenue
Detroit, MI 48203
(313) 366-3500

Swanson-Erie Corp.
814 E. 8th St.
P.O. Box 1217
Erie, PA 16512
(814) 453-5841

Thermwood Corporation, Inc.
P.O. Box 436
Acton, MA 01720
(812) 937-4476

Unimation/Westinghouse, Inc.
Shelter Rock Lane
Danbury, CT 06810
(203) 744-1800

Westinghouse Electric Corp.
Industry Automation Div.
400 High Tower Office Building
400 Media Dr.
Pittsburgh, PA 15205
(412) 778-4349

Yaskawa Electric America, Inc.
305 Era Drive
Northbrook, IL 60062
(312) 564-0770

Robot Manufacturers (Europe)

KUKA
Schweissanlagen & Roboter GmbH
P.O. Box 431280
Zugspitzstr. 140

D-8900
Augsburg 43
West Germany

Nimak
Werkstrabe
Postfach 86
5248 Wissen/Sieg
West Germany

Volkswagenwerk AG
Abt. Industrieverkauf
3180 Wolfsburg
West Germany

Electrolux AB
Industrial Systems
S-105 45 Stockholm
Sweden

Jungheinrich
    Unternelmensverwaltung
Friedrich-Ebert-Dabb 129
2000 Hamburg 70
West Germany

ASEA AB
S-72183 Vasteras
Sweden

R. Kaufeldt AB
P.O. Box 42139
S-126 Stockholm
Sweden

British Federal Welder &
    Machine Co., Ltd.
Castle Mill Works
Dudley
West Midlands, DY1 4DA
United Kingdom

Hall Automation Limited
Colonia Way
Watford
Herts, WD2 4FG
United Kingdom

Mouldmation Limited
2 Darwin Close
Burntwood, Walsall
Staffs WS7 9HP
United Kingdom

Pendar
Bridgwater
Somerset
United Kingdom

Unimation, Inc.
Units A3/A4
Stafford Park 4
Telford, Salop
United Kingdom

A.O.I.P. Kremlin Robotique
6 rue Maryse Bastie
9100 Evry
France

Regie Nationale des
Usines renault SA
66Av Edouard Vaillaut
Boulogne-Billancourt
France

Sormel
rue Becquerel
25009 Besanicon Cedex
France

Camel Robot SRL
Palozzolo Milanese
Italy

Digital Electronics Automation
Co Torino 70
Moncalieri, Piemonte 10024
Italy

Fiat Auto S.p.A.
CSO Agnelli 200
Torino, Piemonte
Italy

Olivetti S.p.A.
Controllo Numerico
Fr S Bernardo
V Torino 603
Ivrea, Piemonte
Italy

Trallfa
Paint-Welding Robot Systems
P.O. Box 113
4341 Bryne
Norway

Robot Manufacturers (Japan)

Dainichi Kiko Co., Ltd.
Kosai-cho
Nakakomagun Yeamanshi Pref.
400-04
Japan

Fanuc, Ltd.
3–5–1
Asahigoaka, Hino City
Tokyo
Japan

Hitachi, Ltd.
Shin-Maru Bldg.
1–5–1
Marunouchi, Chiyoda-ku
Tokyo
Japan

Kawasaki Heavy Industries Ltd.
World Trade Center Bldg.
2–4–1
Hamamatsucho, Minato-ku
Tokyo
Japan

Matsushita Industrial
  Equipment Co. Ltd.
3–1–1 Inazumachi
Toyonaka City Osaka Pref.
Japan

Mitsubishi Heavy Industries Ltd.
2–5–1
Marunouchi, Chiyoda-ku
Tokyo
Japan

Sankyo Seiki Mfg. Co., Ltd.
1–17–2
Shunbashi, Minati-ku
Tokyo 105
Japan

Tokico Ltd.
1–6–3
Funta, Kawasaki-ku
Kawasaki City
Kanagaw Pref.
Japan

Yaskawa Electric Mfg. Co. Ltd.
Ohtemachi Bldg.
1–6–1
Ohtemachi, Chiyoda-ku
Tokyo
Japan

Robotics Associations

Robot Institute of America
One SME Drive
P.O. Box 930
Dearborn, MI 48128
(313) 271-0778

Robotics International of SME
One SME Drive
P.O. Box 930
Dearborn, MI 48128

Japan Industrial Robot
  Association (JIRA)
3-5-8 Shiba-koen
Minato-ku
Tokyo 105
Japan

British Robot Association
35-39 High Street
Kempston
Bedford MK42 7BT
England

Swedish Industrial Robot
  Association (SWIRA)
Box 5506
Storgatan 19
S-14 85 Stockholm
Sweden

Association Francaise de Robotique
  Industrieele (AFRI)
89 Rue Falgueire
75015 Paris,
France

Societa Italiana Robotica
  Industriale (SIRI)
Instituto di Elettrotechnica
  ed Elettronics
Politechmico di Milano
Piazza Leonardo da Vinci 32
20133 Milano
Italy

Robot Consulting/Applications Firms

Automation Systems/American
  Technologies
1900 Pollitt Drive
Fair Lawn, NJ 07410
(201) 797-8200

Blanarovich Engineering
Box 292
Don Mills, Ontario M3C 2S2
Canada
(416) 438-6313

RMT Engineering Ltd.
P.O. Box 2333, Station B
St. Catherines, Ontario L2M 7M7
Canada
(416) 937-1550

Robot Systems, Inc.
50 Technology Parkway
Norcross, GA 30092
(404) 448-4133

Technology Research Corporation
8328-A Traford Lane
Springfield, VA 22152
(703) 451-8830

Machine Vision System Organizations
System Manufacturers

Adaptive Technologies, Inc.
600 W. North Market Blvd., #1
Sacramento, CA 95834
(916) 920-9119

Adept Technology, Inc.
1212 Bordeaux Drive
Sunnyvale, CA 94089
(408) 747-0111

Analog Devices
3 Technology Way
Norwood, MA 02062
(617) 329-4700

Applied Intelligent Systems
110 Parkland Plaza
Ann Arbor, MI 48103
(313) 995-2035

Applied Scanning Technology
1988 Leghorn St.
Mountain View, CA 94043
(415) 967-4211

ASEA Robitics Inc.
16250 W. Glendale Dr.
New Berlin, WI 53151
(414) 875-3400

Autoflex Inc.
25880 Commerce Drive
Madison Heights, MI 48071
(313) 398-9911

Automatic Inspection Devices
One SeaGate
Toledo, OH 43666
(419) 247-5000

Automatic Vision
1300 Richard St.
Vancouver, BC V6B 3G6
CN

Automation Intelligence
1200 W. Colonial Dr.
Orlando, FL 32804-7194
(305) 237-7030

Automation Systems Inc.
1106 Federal Rd.
Brookfield, CT 06804
(203) 775-2581

Automatix
1000 Tech Park Dr.
Billerica, MA 01821
(617) 667-7900

Beeco Inc.
4175 Millersville Rd.
Indianapolis, In 46205
(317) 547-1717

Cambridge Instruments Inc.
    Sub. Cambridge Instruments Ltd.
40 Robert Pitt Dr.
Monsey, NY 10952
(914) 356-3331 or 356-3877

CBIT Corporation
Horse Shoe Trail RD #2
Chester Springs, PA 19425
(215) 469-0358

Cognex
72 River Park St.
Needham, MA 02194
(617) 449-6030

Computer Systems Co.
26401 Harper Ave.
St. Clair Shores, MI 48081
(313) 779-8700

Contrex
47 Manning Rd.
Billerica, MA 01821
(617) 273-3434

Control Automation
Princeton-Windsor Industrial Park
P.O. Box 2304
Princeton, NJ 08540
(609) 799-6026

COSMOS Imaging Systems, Inc.
30100 Crown Valley
    Parkway Suite 32
Laguna Niguel, CA 92677
(714) 495-2662

Cybotech
P.O. Box 88514
Indianapolis, IN 46208
(317) 298-5890

DCI Corp.
110 S. Gold Dr.
Robbinsville, NJ 08691
(609) 587-9132

Diffracto, Ltd.
6360 Hawthorne Dr.
Windsor, Ontario N8T 1J9
Canada
(519) 945-6373

Digital/Analog Design Assoc.
530 Broadway
New York, NY 10012
(212) 966-0410

Eaton Corporation
4201 N. 27th St.
Milwaukee, WI 53216
(414) 449-6345

Eigen/Optivision
P.O. Box 848
Nevada City, CA 95959
(916) 272-3461

Electro-Optical Information Systems
710 Wilshire Blvd.
Suite 501
Santa Monica, CA 90401
(213) 451-8566

Everett/Charles Test Equipment, Inc.
2887 N. Towne Ave.
Pomona, CA 91767
(714) 621-9511

Federal Products Corp.
Boice Division
P.O. Box 12-185
Albany, NY 12212
(518) 785-2211

Gallaher Enterprises
P.O. Box 10244
Winston-Salem, NC 27108-0318
(919) 725-8494

General Electric
P.O. Box 17500
Orlando, FL 32860-7500
(305) 889-1200

General Numeric Corp.
390 Kent Ave.
Elk Grove Fillage, IL 60007
(312) 640-1595

GMF Robotics Corp.
Northfield Hills Corp. Center
5600 New King St.
Troy, MI 48098
(313) 641-4242

Hamamatsu Systems
40 Bear Hill Rd.
Waltham, MA 02154
(617) 890-3440

Ham Industries
835 Highland Rd.
Macedonia, OH 44056
(216) 467-4256

Hitachi America, Ltd.
50 Prospect Ave.
Tarrytown, NY 10591-4698
(914) 332-5800

Honeywell Visitronics
P.O. Box 5077
Englewood, CO 80155
(303) 850-5050

Image Data Systems
315 W. Huron, Suite 140
Ann Arbor, MI 48103
(313) 761-7222

Industrial Technology and
   Machine Intelligence
1 Speen St., Suite 240
Framingham, MA 01701
(617) 620-0184

Industrial Vision Systems, Inc.
Victory Research Park
452 Chelmsford St.
Lowell, MA 01851
(617) 459-9000

Integrated Automation
2121 Allston Way
Berkeley, CA 94704
(415) 843-8227

Intelledex Inc.
33840 Eastgate Circle
Corvallis, OR 97333
(503) 758-4700

International Imaging Systems
1500 Buckeye Dr.
Milpitas, CA 95035
(408) 262-4444

International Robomation/
   Intelligence
2281 Las Palmas Dr.
Carlsbad, CA 92008
(619) 438-4424

IRT Corp.
3030 Callan Rd.
P.O. Box 85317
San Diego, CA 92138-5317
(619) 450-4343

Itek Optical Systems/Litton
   Industries
10 Maguire Rd.
Lexington, MA 02173
(617) 276-2000

ITRAN
670 N. Commercial St.
P.O. Box 607
Manchester, NH 03105
(603) 669-6332

Key Image Systems
20100 Plummer St.
Chatsworth, CA 91311
(213) 993-1911

KLA Instruments
2051 Mission College
Santa Clara, CA 95054
(408) 988-6100

Machine Intelligence Corp.
330 Potrero Ave.
Sunnyvale, CA 94086
(408) 737-7960

Machine Vision International
Burlington Center
325 Eisenhower
Ann Arbor, MI 48104
(313) 996-8033

Mack Corporation
3695 East Industrial Dr.
P.O. Box 1756
Flagstaff, AZ 86002
(602) 526-1120

Micro-Poise
P.O. Box 88512
Indianapolis, IN 46208
(317) 298-5000

Nikon Instrument Group
623 Stewart Ave.
Garden City, NY 11530

Object Recognition Systems
1101-8 State Rd.
Princeton, NJ 08540
(609) 924-1667

Octek
7 Corporate Place
South Bedford St.
Burlington, MA 01803
(617) 273-0851

Opcon
720 80th St. S.W.
Everett, WA 98203
(206) 353-0900

Optical Gaging Products
850 Hudson Ave.
Rochester, NY 14621
(716) 544-0400

Optical Specialties
4281 Technology Drive
Fremont, CA 94538
(415) 490-6400

Optrotech Inc.
Suite 206
111 S. Bedford St.
Burlington, MA 01803
(617) 272-4050

Pattern Processing Technologies
511 Eleventh Ave. S.
Minneapolis, MN 55415
(612) 339-8488

Perceptron
23855 Research Dr.
Farmington Hills, MI 48024
(313) 478-7710

Photonic Automation, Inc.
3633 W. McArthur Blvd.
Santa Ana, CA 92704
(714) 546-6651

Photo Research Vision Systems
Division of Kollmorgen
3099 N. Lima St.
Burbank, CA 91504
(818) 954-0104

Prothon
Div. of Video Tek
199 Pomeroy Rd.
Parsippany, NJ 07054
(201) 887-8211

Quantex Corp.
252 N. Wolfe Rd.
Sunnyvale, CA 94086
(408) 733-6730

Rank Videometrix
9421 Winnetka
Chatsworth, CA 91311
(818) 343-3120

Recognition Concepts Inc.
924 Incline Way
P.O. Box 8510
Incline Village, NV 89450
(702) 831-0473

Robotic Vision Systems
425 Rabro Dr. E.
Hauppauge, NY 11788
(526) 273-9700

Selcom
P.O. Box 250
Valdese, NC 28690
(704) 874-4102

SILMA, Inc.
1800 Embarcadero
Palo Alto, CA 94301
(415) 493-0145

Spatial Data Systems
420 S. Fairview Ave.
P.O. Box 978
Goleta, CA 93117
(805) 967-2383

Syn-Optics
1225 Elko Dr.
Sunnyvale, CA 94089
(408) 734-8563

Synthetic Vision Systems
2929 Plymouth Rd.
Ann Arbor, MI 48105
(313) 665-1850

Technical Arts Corp.
180 Nickerson St. #303
Seattle, WA 98109
(206) 282-1703

Testerion, Inc.
9645 Arrow Hwy.
P.O. Box 694
Cucamonga, CA 91730
(714) 987-0025

Time Engineering
1630 Big Beaver Rd.
Troy, MI 48083
(313) 528-9000

Unimation
Shelter Rock Lane
Danbury, Ct 06810
(203) 744-1800

United Detector Technology
12525 Chadron Ave.
Hawthorne, CA 90250
(213) 978-0516

Vektronics
5750 El Camino Real
P.O. Box 459
Carlsbad, CA 92008
(619) 438-0992

Vicom Systems
2520 Junction Ave.
San Jose, CA 95134
(408) 946-5660

Videk
Division of Eastman Technology
343 State St.
Rochester, NY 14650
(800) 445-6325, Ext. 15

View Engineering
1650 N. Voyager Ave.
Simi Valley, CA 93063
(805) 522-8439

Visionetics
P.O. Box 189
Brookfield Center, CT 06805
(203) 775-4770

Vision System Technologies
1532 S. Washington Ave.
Piscataway, NJ 08854
(201) 752-6700

Visual Matic
2171 El Camino Real
Oceanside, CA 92054
(619) 722-8299

Vuebotics Corp.
6086 Corte Del Cedro
Carlsbad, CA 92008
(619) 438-7994

3M, Vision Systems
Suite 300
8301 Greensboro Dr.
McLean, VA 22102
(703) 734-0300

Machine Vision Consultants and System Integrators

Anorad Corp.
110 Oser Ave.
Hauppauge, NY 11788
(516) 231-1990

Automated Vision Systems
1590 La Pradera Dr.
Campbell, Ca 95008
(408) 370-0229

Automation Unlimited
10 Roessler Rd.
Woburn, MA 01801
(617) 933-7288

Bahr Technologies Inc.
1842 Hoffman St.
Madison, WI 53704
(608) 244-0500

Hansford Manufacturing
3111 Winton Road South
Rochester, NY 14623
(716) 427-8150

Image Technology Methods Corp.
103 Moody St.
Waltham, MA 02154
(617) 894-1720

Key Technology Inc.
517 N. Elizabeth
P.O. Box 8
Milton-Freewater, OR 97862

Medar, Inc.
38700 Grand River Ave.
Farmington Hills, MI 48018
(313) 477-3900

Multicon
7035 Main St.
Cincinnati, OH 45244
(513) 271-0200

Raycon Corp.
77 Enterprise
Ann Arbor, MI 48103
(313) 769-2614

Robotic Objectives, Inc.
60 Church St., The Forum
Yalesville, CT 06492
(203) 269-5063

Spectron Engineering
800 W. 9th Ave.
Denver, CO 80204
(303) 623-8987

TAU Corp.
10 Jackson St., Ste. 101
Los Gatos, CA 95030
(408) 395-9191

Technology Research Corporation
8328-A Traford Lane
Springfield, VA 22152
(703) 451-8830

Tech Tran Corporation
134 N. Washington St.
Naperville, IL 60540
(312) 369-9232

Vision Systems International
3 Milton Dr.
Yardley, PA 19067
(215) 736-0994

Visual Intelligence Corp.
Amherst Fields Research Park
160 Old Farm Road
Amherst, MA 01002
(413) 253-3482

Research Organizations

Carnegie-Mellon University
The Robotics Institute
Pittsburgh, PA 15213
(412) 578-3826

Environmental Research Institute
of Michigan
Robotics Program
P.O. Box 8618
Ann Arbor, MI 48107
(313) 994-1200

George Washington University
725 23rd Street, N.W.
Washington, DC 20052
(202) 676-6919

Jet Propulsion Labs
Robotics Group
4800 Oak Grove Drive
Passadena, CA 91103
(213) 354-6101

Massachusetts Institute of
  Technology
Artificial Intelligence Lab
545 Technology Square
Cambridge, MA 02139
(617) 253-6218

National Bureau of Standards
Industrial Systems Division
Building 220, Room A123
Washington, DC 20234
(301) 921-2381

Naval Research Lab
Code 2610
Washington, DC 20375
(202) 767-3984

North Carolina State University
Department of Electrical
  Engineering
Raleigh, NC 27650
(919) 737-2376

Purdue University
School of Electrical Engineering
West Lafayette, IN 47906
(317) 749-2607

Rensselaer Polytechnic Institute
Center for Manufacturing
  Productivity
Jonsson Engineering Center
Troy, NY 12181
(518) 270-6724

SRI International
Artificial Intelligence Center
Menlo Park, CA 94025
(425) 497-2797

Stanford University
Artificial Intelligence Lab
Stanford, CA 94022
(415) 497-2797

U.S. Air Force
AFWAL/MLTC
Wright Patterson AFB, OH 45437
(513) 255-6976

University of Central Florida
IEMS Department
Orlando, FL 32816
(305) 275-2236

University of Cincinnati
Institute of Applied
  Interdisciplinary Research
Location 42
Cincinnati, OH 45221
(513) 475-6131

University of Maryland
Computer Vision Laboratory
College Park, MD 20742
(301) 454-4526

University of Rhode Island
Department of Electrical
  Engineering
Kingston, RI 02881
(401) 792-2187

University of Southern California
School of Engineering
University Park
Los Angeles, CA

University of Texas
Austin, TX 78712
(512) 471-1331

University of Washington
Department of Electrical
  Engineering
Seattle, WA 98195
(206) 543-2056

Glossary

ASSEMBLY ROBOT: A robot designed, programmed, or dedicated to putting together parts into subassemblies or complete products.

AUTOMATED STORAGE AND RETRIEVAL SYSTEM—AS/RS: A high density rack storage system with rail running vehicles serving the rack structure for automatic loading and unloading of material.

AUTOMATION: The theory, art, or technique of making a process automatic, self-moving or self-controlling.

BATCH: A group of workpieces processed at one time.

BATCH MANUFACTURING: The production of parts in discrete runs or batches, interspersed with other production operations or runs of other parts.

BUFFER: A temporary holding area. In an automated cell or workstation, a buffer area might be used to store incoming semi-finished metal until the machine tools are ready to process them, or to store finished parts until the materials handling system is ready to fetch them. In a computer, a buffer is an area of memory set aside to hold data prior to being moved from one part of the system to another.

BURN-IN: A product test where all components are operated continuously for an extended period of time to eliminate early failures.

CELL: A manufacturing unit consisting of two or more workstations with the material handling systems, storage buffers, and other items necessary to connect them.

CENTRAL PROCESSING UNIT (CPU): Mainframe, or, more specifically, the central processor of a computer system, containing the arithmetic unit and logic. (In a microcomputer, the CPU is often on a single chip called a microprocessor.)

COLLET: A split cylindrical sleeve that holds a tool or workpiece on the drive shaft of a lathe or other machine.

COMPUTER-AIDED DESIGN (CAD): The use of a computer and computer graphics in the design of parts, products, etc. Some advanced CAD systems not only produce three-dimensional views of intricate parts, but also allow the designer to "test" a simulated part under different stresses, loads, and so forth.

COMPUTER-AIDED MANUFACTURING (CAM): The use of computers in the management, control, and operation of a manufacturing plant.

COMPUTER-INTEGRATED MANUFACTURING (CIM): The concept of a totally automated factory in which all manufacturing processes are integrated and controlled by computer. CIM enables production planners and schedulers, shop-floor foremen, and accountants to use the same database as product designers and engineers.

COMPUTER NUMERICAL CONTROL (CNC): A technique in which a machine tool control uses a minicomputer to store numerical control instructions generated earlier by CAD/CAM for controlling the machine.

CONTROLLER: A computer, or group of computers, used to control a machine tool, robot, or similar device or group of devices. They may be arranged in a hierarchy so that, for example, a workstation controller may issue "commands" to a robot controller or machine tool controller.

COORDINATE MEASURING MACHINE (CMM): A machine for measuring the shape and dimensions of small, solid objects, particularly objects with a complex shape. Usually it has a small probe which is touched against or traced around the surface of the object to be measured. The machine reports the position of the tip of the probe from point to point in three-dimensional coordinates. The dimensions of the object can then be calculated from these coordinates. The entire operation can be automated.

DIRECT NUMERICAL CONTROL (DNC): The use of a shared computer to distribute part program data by way of data lines to remote machine tools.

DIRTY POWER: Vague term used to describe electric power that is sufficiently adulterated with harmonic distortion and other impurities to cause electricity using machinery to operate less than optimally.

DISCRETE PART or BATCH MANUFACTURING: A manufacturing process that produces discrete parts (as opposed to continuous-process industries, such as oil refining or paper making) in comparatively small lots, or batches, of from 1 to perhaps 50,000.

ELECTRICAL DISCHARGE MACHINING: The workpiece is precisely eroded or cut by electric pulses jumping between an electrode and the workpiece in the presence of a dielectric fluid.

EMULATION: In computer science, a program which allows one machine to execute instructions meant for a different machine, as opposed to a simulation which merely allows the first machine to act like a second machine.

END EFFECTOR: A tool or gripping mechanism attached to the "wrist" of a robot to accomplish some task. While gripping mechanisms can be thought of as robotic "hands," end effectors also include single-purpose attachments such as paint guns, drills, and arc-welders.

FIXTURE: A device which holds a workpiece in position in a machine tool for machining. The workpiece must be held in a precise position, with no room for slippage, so if the shape is at all complex, a special fixture is usually built to hold the piece. One challenge in the automation of small-batch manufacturing is the development of flexible fixturing systems that can adapt to a wide variety of differently sized and shaped workpieces.

FLEXIBLE: Of manufacturing systems or facilities: easily adapted to produce new products; easily redirected or reprogrammed.

FLEXIBLE AUTOMATION: Refers to the multi-task capability of robots; multi-purpose, adaptable, reprogrammable.

FLEXIBLE MACHINING CENTER (FMC): Usually a multi-robot system that comprises CNC turning or grinding machines with robots loading and unloading parts that are conveyed into and through the system.

FLEXIBLE MANUFACTURING SYSTEM (FMS): A group of NC machine tools that can randomly process a group of parts having different process sequences and process cycles using automated material handling and central computer control to dynamically balance resource utilization so that the system can adapt automatically to changes in part production mixes and levels of output.

HARMONIC DISTORTION: Nonlinear distortion of a system or transducer characterized by the appearance in the output of harmonics other than the fundamental component when the input wave is sinusoidal.

HORIZONTAL MACHINING CENTER: The horizontal machining center is primarily for working on cube-shaped parts. Using a cutting tool rotating on a horizontal axis, it can drill holes, mill out pockets, cut grooves, or plane metal from a surface.

HIERARCHICAL CONTROL SYSTEM: A computer control scheme in which the data processing and computation necessary to accomplish a task is split into discrete levels, with the outputs of higher levels being used as input commands for lower levels, and lower levels furnishing status reports as data for higher levels. Each level in the hierarchy accepts tasks from the level above it and splits those tasks into subtasks that are parceled out to the levels below it.

INTERFACE: Literally, a shared boundary. An interface between two computer systems, such as intelligent machine controllers, involves a method for passing commands, responses, and data from one system to another. When the two computers are not inherently compatible, the interface becomes a "translator" between the two systems. An interface can also have a physical component—the proper set of connectors, voltage levels, and so forth necessary to hook two systems together. The common three-prong electrical plug is an example of a standardized interface used in residential electric power systems.

LOAD CYCLE TIME: Time in manufacturing process required for unloading the last workpiece and loading the next one.

LOAD FACTOR: The ratio of the average load over a designated period of time to the peak load occurring in that period.

LOT: Batch or group of workpieces processed at one time.

MACHINING CENTER: A machine capable of performing a variety of metal removal operations on a part, usually under numerical control.

MICROPROCESSOR: A basic element of a central processing unit that is a single integrated circuit. A microprocessor requires additional circuits to become a suitable central processing unit.

MILLING MACHINE: Milling machines use tools such as drills and routers which rotate and have a cutting edge on their periphery. Originally, the term "milling" implied that the rotating tool is held stationary and the workpiece is moved against it to effect the cutting or routing. In many modern milling machines, the tool can also move in and out along one or more axes.

MINICOMPUTER: A class of computer in which the basic element of the central processing unit is constructed of a number of discrete components and integrated circuits rather than being comprised of a single integrated circuit, as in the microprocessor.

NON-SERVO ROBOT: A robot that can be programmed by its motions only to the end points of each of its axes of motion.

NUMERICAL CONTROL (NC): A technique for the control of machine tools or factory processes in which information on the desired actions of the system is prerecorded in numerical form and used to control the operation automatically.

OPERATING SYSTEM: A more or less general-purpose computer program which controls the execution of the specific "applications" programs run on a computer. An operating system provides the critical interface between the computer hardware and the humans that use it, handling things like program scheduling, the "input" and "output" of data, detecting and diagnosing errors in the performance of application programs, and other housekeeping chores.

PAYLOAD: Maximum weight carried at normal speed—also called workload.

POWER FACTOR: The ratio of total watts to the total root-mean-square (RMS) volt-amperes.

PROCESS PLANNING: Specifying the sequence of production steps from start to finish, and describing the state of the workpiece at each workstation. Recently CAM capabilities have been applied to the task of preparing process plans for the fabrication or assembly of parts.

REAL TIME: Refers to tasks or functions executed so rapidly by a CAD/CAM system that the feedback at various stages in the process can be used to guide the user completing the task.

ROBOT: A reprogrammable multifunctional manipulator designed to move material, parts, tools, or specialized devices, through variable programmed motions for the performance of a variety of tasks.

SENSOR: A "transducer" which takes in information about the physical state of things and converts it to an electrical signal that can be processed by a control system. Sensors can be simple, such as a temperature monitor on a machine tool, or highly complex, such as a machine vision system. Other sensors monitor things like torque on a robot wrist, the pressure exerted by robot grippers, or the vibrations in a workpiece being machined.

TURNING CENTER: A lathe, that is, a machine in which the piece being worked rotates on an axis so that tools can work on it. When a lathe becomes sufficiently "intelligent," with the addition of programmable numerical controls, it becomes a "turning center."

VERTICAL MACHINING CENTER: The vertical machining center is similar to the horizontal machining center in that it consists of a robot and a milling machine working together. The vertical machine center differs in that the cutting tool rotates on a vertical axis and can only be moved vertically (Y axis). It can drill or cut only on the top surface of the workpiece.

WORK ENVELOPE: The set of points representing the maximum extent or reach of the robot hand or working tool in all directions.

WORKPIECE: The piece of metal which is being worked.

WORKSTATION: A manufacturing unit generally consisting of one numerical control machine tool or measuring machine, usually tended by a robot.

Bibliography

Books

Ballard, D.H. and C.M. Brown. *Computer Vision*. NJ: Prentice-Hall, 1982.

Baranson, Jack. *Automated Manufacturing: The Key to International Competitiveness—and Why the United States Is Falling Behind*. Washington, D.C.: Developing World Industry & Technology, 1983.

Baxes, G.A. *Digital Image Processing*. NJ: Prentice-Hall, 1984.

Brown, Robert Goodell. *Materials Management Systems*. NY: John Wiley & Sons, 1977.

Considine, Douglas M. *Standard Handbook of Industrial Automation*. NY: Chapman & Hall, 1986.

Dorf, Richard C. *Robotics and Automated Manufacturing*. Reston, VA: Reston Publishing Company, 1983.

Engelberger, J.F. *Robotics in Practice*. Amacom Division of American Management Associations, 1980.

Gomersall, A. *Machine Intelligence: An International Bibliography with Abstracts on Sensors in Automated Manufacturing*. Springer-Verlag, 1984.

Harrington, Joseph, Jr. *Computer Integrated Manufacturing*. Huntington, NY: Robert E. Krieger Publishing Company, 1979.

Hollingum, J. *Machine Vision: The Eyes of Automation*. Springer-Verlag, 1984.

Horn, Berthold Klaus Paul. *Robot Vision*. Cambridge, MA: MIT Press, 1986.

Hunt, V. Daniel. *Industrial Robotics Handbook*. NY: Industrial Press, 1983.

———. *Smart Robots*. NY: Chapman and Hall, 1985.

———. *Artificial Intelligence and Expert Systems Sourcebook*. NY: Chapman and Hall, 1986.

———. *Robotics Sourcebook*. NY: Elsevier, 1987.

———. *Mechatronics—Japan's Newest Threat*. NY: Chapman and Hall, 1987.

————. *Dictionary of Advanced Manufacturing Technology.* NY: Elsevier, 1987.

Lambert, Douglas M. *The Development of an Inventory Costing Methdology: A Study of the Costs Associated with Holding Inventory.* Chicago: National Council of Physical Distribution Management, 1975.

Lester, Ronald H., Enrick, Norbert L., Mottley, Jr., Harry E. *Quality Control for Profit.* NY: Industrial Press, 1977.

Makridakis, Spyros, and Wheelright, Steven C. *Forecasting—Methods and Applications.* NY: John Wiley & Sons, 1978.

Minsky, Marvin. *Robotics.* Garden City, NY: Anchor Press/Doubleday, 1985.

Orlicky, Joseph. *Material Requirements Planning.* NY: McGraw-Hill, 1975.

Plossl, George W. *Manufacturing Control—The Last Frontier for Profts.* Reston, VA: Reston Publishing Company, 1973.

Plossl, George W., and Welch, W. Evert. *The Role of Top Management in the Control of Inventory.* Reston, VA: Reston Publishing Company, 1979.

Smith, Bernard T. *Focus Forecasting—Computer Techniques for Inventory Control.* Boston: CBI Publishing Company, 1978.

Susnjara, Ken. "A Manager's Guide to Industrial Robots," Corinthian Press, 1982.

Wight, Oliver. *Production and Inventory Management in the Computer Age.* Boston: Cahners Books International, 1974.

Reports

Aron, Paul H. *The Robot Scene in Japan: An Update.* September 7, 1983, pp. 40–42.

Competitive Assessment of the U.S. Metalworking Machine Tool Industry. Washington, D.C.: United States International Industry. Washington, D.C.: United States International Trade Commission, September 1983.

Competitive Position of U.S. Producers of Robotics in Domestic and World Markets. Washington, D.C.: United States International Trade Commission, December 1984.

Computerized Manufacturing Automation: Employment, Education, and the Workplace. Washington, D.C.: U.S. Congress, Office of Technology Assessment, April 1984.

Conigliaro, Laura and Christine Chien. "CIM Newsletter." Prudential Bache Securities.

Country Market Survey: Machine Tools—United Kingdom. CMS 79-11: Washington, D.C.: U.S. Department of Commerce, Industry and Trade Administration, September 11, 1979, p. 4.

Department of Commerce, International Trade Administration. "A Competitive Assessment of the U.S. Computer-Aided Design and Manufacturing Systems Industry." February 1987.

———. "A Competitive Assessment of the U.S. Flexible Manufacturing Systems Industry." July 1985.

———. "A Competitive Assessment of the U.S. Robotics Industry." March 1987.

Department of Commerce, Office of Industry Assessment, Industry Analysis Division. "A Competitive Assessment of the U.S. Manufacturing Automation Equipment Industries." U.S. Department of Commerce technical publication, June 1984.

Dun & Bradstreet, Inc. "An Analysis of the Robotics Industry." 1983.

Fisk, J.D. "Industrial Robots in the United States: Issues and Perspectives." Congressional Research Service, The Library of Congress, Report No. 81-78 E, March 30, 1981.

Flexible Manufacturing System Handbook, volumes 1-4. Charles Stark Draper Laboratory, Cambridge, Mass., February 1983.

"Flexible Manufacturing Systems—State of the Art and Trends in Their Diffusion." Economic Commission for Europe, Working Party on Engineering Industries and Automation, July 1984.

Flexible MFG System-Based Factory Automation in Japan. Tokyo, Japan: Yano Research Institute Ltd., December 1983.

Foreign Industrial Targeting and its Effects on U.S. Industries Phase I: Japan. Washington, D.C.: United States International Trade Commission, October 1983.

Gevarter, William B. "An Overview of Artificial Intelligence and Robotics: Volume 2—Robotics." U.S. Department of Commerce, National Bureau of Standards, March 1982.

———. "An Overview of Computer Vision," U.S. Department of Commerce, National Bureau of Standards, September 1982.

"Industrial Robots—Forecasts and Trends." Delphi Study, 2nd ed., Society of Manufacturing Engineers/The University of Michigan, 1985.

Industrial Robots: Their Role in Manufacturing Industry. Paris, France: Organization for Economic Cooperation and Development, 1983.

Industrial Targeting—Foreign Industrial Targeting Practices and Possible U.S. Responses. A Report to the President from the Presidents Export Council, December 1984.

Machine Tools & Accessories Market in Far East. Frost & Sullivan Inc., March 1981, p. 104.

National Electrical Manufacturers Association (NEMA). "U.S. Manufacturers Five-Year Industrial Automation Plans for Automation Machinery and Plant Communication Systems." March 15, 1985.

National Research Council. *The U.S. Machine Tool Industry and the Defense Industrial Base.* Washington, D.C.: National Academy Press, 1983.

U.S. Congress, Office of Technology Assessment. "Computerized Manufacturing Automation: Employment, Education, and the Workplace." U.S. Government Printing Office, OTA–CIT–235, April 1984.

Directories

CAD/CAM, CAE: Survey, Review
  and Buyers Guide.
DARATECH, Inc.
Cambridge, MA 02138

CAD/CAM; Robotics; and
  Industrial Sensor; Industry
  Directory.
Techical Database Corporation
Conroe, TX

Industrial Robots.
Society of Manufacturing
  Engineers
Marketing Services Division
One SME Drive
P.O. Box 930
Dearborn, MI 48128

Periodicals

Assembly Engineering
Hitchcock Publishing Company
New York, NY

Automation News
The Newspaper for Factory
  Automation
Grant Publications, Inc.
New York, NY

CIM
CADLINC Inc.
Elk Grove Village, IL

Electronic Imaging
Morgan-Grampian Publishing Co.
Berkshire Common
Pittsfield, MA 01201
(413) 499-2550

Image and Vision Computing
Butterworth Scientific Ltd.
P.O. Box 63
Westbury House, Bury St.
Guildford, Surrey GU1 5BH
United Kingdom

Industrial Robots International
158 Linwood Plaza
P.O. Box 13304
Fort Lee, NJ 07024

Le nouvel Automatisme
41 rue de la Grange-aux Belles
75483 Paris Cedex 10
France

Managing Automation
Thomas Publishing Company
New York, NY

Manufacturing Engineering
Society of Manufacturing Engineers
Dearborn, Michigan

Manufacturing Technology Horizons
Tech Tran Corporation
134 N. Washington Street
Naperville, IL 60540
(312) 369-9232

Material Handling Engineering
Penton/IPC
Cleveland, OH

Photonics Spectra
Optical Publishing Co.
P.O. Box 1146
Berkshire Common
Pittsfield, MA 01202

Production Engineering
Penton/IPC
Cleveland, OH

Robomatics Reporter
EIC/Intelligence
38 W. 38th Street
New York, NY 10018

Robot Insider
Fairchild Publications
7 E. 12th Street
New York, NY 10003

Robot News International
IFS (Publications) Ltd.
35–39 High Street
Kempston, Bedford
MK42 7BT England

Robot/X News
Robotics Publications
P.O. Box 450
Mansfield, MA 02048

Robotech Japan
Topika Inc.
Nagatani Bldg.
7-17-4, Ginza
Chuko-ku, Tokyo 104
Japan

Robotics Age
P.O. Box 725
La Canada, CA 91011

Robotics Technology Abstracts
Tech Tran Corporation
134 N. Washington Street
Naperville, IL 60540
(312) 369-9232

Robotics Technology Abstracts
Cranfield Press
Management Library
Cranfield Institute of Technology
Cranfield Bedford MK43 OAL
United Kingdom

Robotics Today
Society of Manufacturing Engineers
One SME Drive
P.O. Box 930
Dearborn, MI 48128

Robotics World
Communication Channels
6255 Barfield Road
Atlanta, GA 30328

Sensors
North American Technology
174 Concord Street
Peterborough, NH 03458
(603) 924-7261

The Industrial Robot
IFS (Publications) Ltd.
35–39 High Street
Kempston, Bedford
MK42 7BT England

The International Journal of
  Robotics Research
The MIT Press
28 Carleton Street
Cambridge, MA 02142

The Robotics Report
Washington National News Reports
Suite 400
7620 Little River Turnpike
Annandale, VA 22003

Vision
Society of Manufacturing Engineers
One SME Drive
P.O. Box 930
Dearborn, MI 48128
(313) 271-1500

Acronyms and Abbreviations

AGVS:	Automated Guided Vehicle System
AI:	Artificial Intelligence
AMH:	Automated Material Handling
AMT:	Advanced Manufacturing Technology
APT:	Automatically Programmed Tools
AS/RS:	Automated Storage and Retrieval System
BOM:	Bill Of Materials
CAD:	Computer-Aided Design
CAD/CAM:	Computer-Aided Design/Computer-Aided Manufacturing
CADD:	Computer-Aided Design and Drafting
CAE:	Computer-Aided Engineering
CAM:	Computer-Aided Manufacturing
CAPP:	Computer-Aided Process Planning
CCD:	Charge Coupled Device
CID:	Charge Injection Device
CIM:	Computer-Integrated Manufacturing
CNC:	Computer Numeric Control
CRT:	Cathode Ray Tube
DDL:	Data Definition Language
DDMIS:	Data Driven Management Information System
DSS:	Decision Support System
DNC:	Direct Numerical Control
ES:	Expert System or Electrical Schematic
EAW:	Equivalent Annual Worth
ECSL:	Extended Control and Simulation Language
FMC:	Flexible Manufacturing Cell, Flexible Manufacturing Center

FMS:	Flexible Manufacturing System
GT:	Group Technology
IAP:	Investment Analysis Program
IC:	Integrated Circuit
IDEF:	ICAM Definition Method, Version Zero
IDSS:	ICAM Definition Support System
IGES:	Initial Graphics Exchange Specification
I/O:	Input/Output
ISO:	International Standards Organization
LAN:	Local Area Network
LAN-WAN:	Local Area Network to Wide Area Network Communications
LCD:	Liquid Crystal Display
LED:	Light Emitting Diode
MAP:	Manufacturing Automation Protocol
MARR:	Minimum Acceptable Rate of Return
MHS:	Material Handling System
MIS:	Management Information System
MRP:	Material Requirements Planning
MRPS:	Manufacturing Resource Planning System
MTBF:	Mean Time Between Failures
MTTR:	Mean Time to Repair
MUM:	Methodology for Unmanned Manufacturing
NC:	Numerical Control
NPV:	Net Present Value
NWP:	Net Present Worth
OSI:	Open System Interconnection
PC:	Personal Computer, Program Counter, or Programmable Controller
PD:	Programmable Device
QA:	Quality Assurance
QC:	Quality Control
RAM:	Random Access Memory
RCC:	Remote Center Compliance
RFP:	Request for Proposal
ROM:	Read Only Memory
SCARA:	Selective Compliance Assembly Robot Arm
SME:	Society of Manufacturing Engineers
TCP:	Tool Center Point
TOP:	Technical and Office Protocols
VLSI:	Very Large Scale Integrated Circuit
VTL:	Vertical Turret Lathes
WAN:	Wide Area Network

INDEX